Pocket Reference to Alzheimer's Disease Management

Pocket Reference to Alzheimer's Disease Management

Anna Burke, MD
Geriatric Psychiatrist,
Dementia Specialist
Banner Alzheimer's Institute
Phoenix, AZ

Geri R Hall, PhD, ARNP, GCNS, FAAN
Advanced Practice Nurse
Banner Alzheimer's Institute
Phoenix, AZ

Roy Yaari, MD, MAS
Behavioral Neurologist, Dementia
Specialist, Associate Director
Banner Alzheimer's Institute
Phoenix, AZ

Adam Fleisher, MD
Behavioral Neurologist
Banner Alzheimer's Institute
Phoenix, AZ

Jan Dougherty, RN, MSN
Director of Family and
Community Services
Banner Alzheimer's Institute
Phoenix, AZ

Jeffrey Young, BA
Psychometrist
Banner Alzheimer's Institute
Phoenix, AZ

Helle Brand, PA
Physician Assistant
Banner Alzheimer's Institute
Phoenix, AZ

Pierre Tariot, MD
Associate Director of Banner
Alzheimer's Institute
Director of the Banner Alzheimer's
Institute Memory Disorders Center
Banner Alzheimer's Institute
Phoenix, AZ

Springer Healthcare

Published by Springer Healthcare Ltd, 236 Gray's Inn Road, London, WC1X 8HB, UK.

www.springerhealthcare.com

© 2015 Springer Healthcare, a part of Springer Science+Business Media.

British Library Cataloguing-in-Publication Data.

A catalogue record for this book is available from the British Library.

ISBN 978-1-910315-21-7

Although every effort has been made to ensure that drug doses and other information are presented accurately in this publication, the ultimate responsibility rests with the prescribing physician. Neither the publisher nor the authors can be held responsible for errors or for any consequences arising from the use of the information contained herein. Any product mentioned in this publication should be used in accordance with the prescribing information prepared by the manufacturers. No claims or endorsements are made for any drug or compound at present under clinical investigation.

Project editor: Laura Hajba

Contents

Author biographies

Anna Burke, MD, is a board certified geriatric psychiatrist and a specialist in dementia care. She has worked extensively with patients and families suffering with Alzheimer's disease (AD) and related dementias. Her area of expertise includes diagnosis and treatment of dementia, as well as treatment of associated behavioral and psychiatric disturbances.

Geraldine R Hall, PhD, ARNP, GCNS, FAAN, is a board-certified clinical nursing specialist who has specialized in care of people affected by dementia since 1980. Her graduate work was completed at the University of Iowa where she taught and became a full professor. Her research and practice is focused on prevention and management of secondary behaviors and helping families to manage. She is widely published and has presented her model, Progressively Lowered Stress Threshold, across the US, Australia, Ireland, and South Korea.

Roy Yaari, MD, MAS, is a board-certified neurologist. He has additional training in a geriatric neurology fellowship and holds a Master's of Advanced Studies in Clinical Research. He has a strong interest in treating dementia patients and in the development of new treatments and therapies.

Adam Fleisher, MD, is a geriatric neurologist, practicing in the Memory Disorders Clinic at the Banner Alzheimer's Institute. He obtained his general neurology training at Johns Hopkins Hospital then completed a clinical and research dementia fellowship at the University of California, San Diego, as well as a Master's degree in Clinical Research. He is an expert in the field of imaging for studying the earliest evidence of AD pathology in the brain, and is well published in the fields of dementia clinical trials and imaging.

Jan Dougherty, RN, MSN, is responsible for setting a new standard of care for patients with dementia and their families through the development and implementation of innovative programs. Jan has extensive experience in dementia care and has developed many pioneering programs being used in Arizona and nationally.

Jeffrey Young, BA, is a clinical psychology doctoral student at the American Professional School of Psychology-Argosy University, Phoenix. He is responsible for neuropsychological testing of patients in the memory disorders clinic. He has a strong interest in early detection and progression of dementia and doing testing research with dementia and cognitively impaired patients.

Helle Brand, PA, has an interest in education and counseling related to dementia, looking at both the effect of dementia on patient's day-to-day functioning and on the extended family. She has experience with dementia as a physician assistant and as a former physical therapist who specialized in the care of older people.

Pierre Tariot, MD, has been Director of the Banner Alzheimer's Institute Memory Disorders Center and Associate Director of the Institute since 2006. A teacher and scholar, he has studied diagnosis and therapy for dementia and Alzheimer's disease, recognition, and management of behavioral disturbances in dementia, and treatment of depression. He has published over 220 papers on these topics, earning awards for his research, such as the American Geriatrics Society New Investigator Award for Neuroscience, an NIMH Geriatric Mental Health Academic Award, and the 2005 UCLA Turken Award. His research affiliations include the NIMH, NIA, Arizona Department of Health, Institute for Mental Health Research, and the Alzheimer's Association.

Abbreviations

Aβ	Amyloid beta
Ach	Acetylcholine
ACHEI	Acetylcholinesterase inhibitors
AD	Alzheimer's disease
ADEAR	Alzheimer's Disease Education and Referral
APP	Amyloid beta precursor protein
ApoE-ε4	Apolipoprotein E-ε4
BPSD	Behavioral and psychological symptoms of dementia
CSF	Cerebrospinal fluid
CT	Computed tomography
DNR	Do not resuscitate
DOT	Department of Transportation
DSM-IV-TR	*Diagnostic and Statistical Manual of Mental Disorders, Fourth Edition, Text Revisions*
EFNS	European Federation of the Neurological Societies
FAD	Familial Alzheimer's disease
FDA	US Food and Drug Administration
FDG-PET	Fluorodeoxyglucose-positron emission tomography
MCI	Mild cognitive impairment
MMSE	Mini-Mental Status Exam
MoCA	Montreal Cognitive Assessment
MRI	Magnetic resonance imaging
NIA	National Institute on Aging
NINDS AD	National Institute on Neurological Disorders and Stroke, Alzheimer's Disease
NMDA	*N*-methyl-D-aspartate
PET	Positron emission tomography
PS-1	Presenilin 1
PS-2	Presenilin 2
RAVLT	Rey Auditory Verbal Learning Test
SPECT	Single photon emission computerized tomography
SPMSQ	Short Portable Mental Status Questionnaire
T-tau	Total tau

Preface

Alzheimer's disease (AD) and related dementias are common conditions and are often diagnosed and treated by primary care providers, yet few have training in state-of-the-art dementia care. Families expect the health care provider to notice when a person loses capacity, diagnose the illness, and manage it both pharmacologically and nonpharmacologically for the remainder of the person's life. It is hoped that this guide will help prepare providers for the complex situations that occur during the diagnoses and management of this terminal disease throughout the disease trajectory to better support the patient, family, and caregivers.

PART ONE

Diagnosis and management of Alzheimer's disease

Chapter 1

Introduction

Epidemiology

The elderly population in the US is expected to grow dramatically due to advances in medicine and the aging generation. As the population ages, the number of Americans with Alzheimer's disease (AD) and other dementias will increase (Figure 1.1) [1,2]. In 2012, an estimated 5.4 million Americans of all ages had AD [3]. Of those with AD, an estimated 4% were under the age of 65 years, 6% were aged 65–74 years, 44% were aged 75–84 years, and 46% were aged 85 years and older [3]. Worldwide, nearly 35.6 million people live with dementia. This number is expected to double by 2030 (65.7 million) and more than triple by 2050 (115.4 million). More than half (58%) of people living with dementia live in low- and middle-income countries. By 2050, this is likely to rise to more than 70% [4].

The estimated annual incidence of AD increases dramatically with age, from approximately 53 new cases per 1,000 people aged 65–74 years, to 170 new cases per 1,000 people aged 75–84 years, to 231 new cases per 1,000 people over the age of 85 years [3]. As the number of people over the age of 65 years in the US increases, the annual total number of new cases of AD and other dementias is projected to double by 2050 (Figure 1.2) [5].

Studies show that people aged 65 years and older generally survive for 4–8 years after the diagnosis of AD, not symptom onset; however, some may live as long as 20 years (Table 1.1) [6–9]. Patients

© Springer Healthcare 2015
A. Burke et al., *Pocket Reference to Alzheimer's Disease Management*,
DOI 10.1007/978-1-910315-22-4_1

with AD will spend more time in the most severe stage of the disease than the early or moderate stages, and will spend much of this stage in a nursing home. 75% of people with AD will be admitted to a

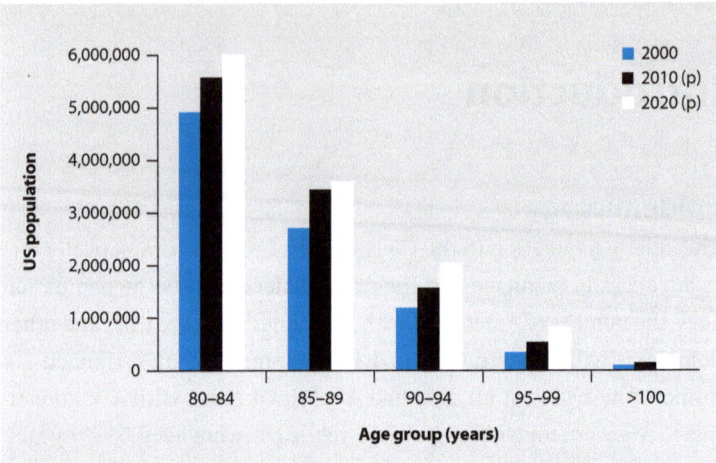

Figure 1.1 Projected number of people in US population with Alzheimer's disease.
United States Census Bureau data and projections illustrate the practical need to understand the correlation between aging and AD. Raw numbers for the US population (2000) and projected (2010 and 2020) are shown. AD, Alzheimer's disease; p, projected. Reproduced with permission from © Springer, 2011. All rights reserved. Nelson et al [2].

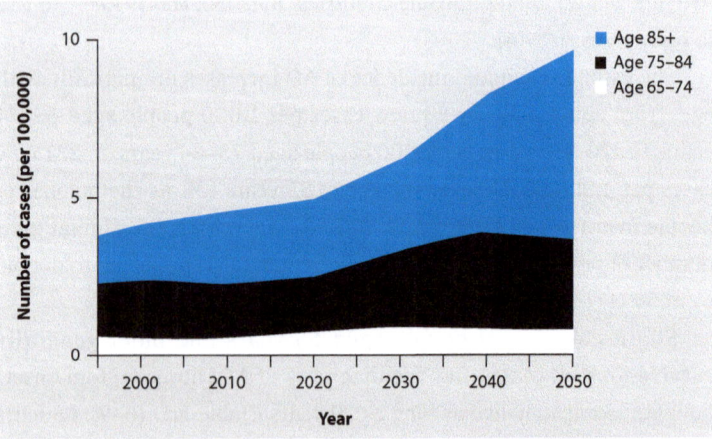

Figure 1.2 Estimated annual number of cases of Alzheimer's disease in the US from 1995 through 2050. Contribution of each of three age groups using the United States Census Bureau middle series population projections. Reproduced with permission from © Lippincott Williams & Wilkins, Inc., 2001. All rights reserved. Herbert et al [5].

nursing home by the age of 80, compared with only 4% of the general population [3,10].

People with dementia have three times as many hospitalizations as older people without dementia [3]. For people with AD and other dementias, aggregate payments for health care, long-term care, and hospice care are projected to increase from $200 billion in 2012 to $1.1 trillion in 2050 [3]. Medicare and Medicaid cover about 70% of the costs of care [3] and, in 2008, average per-person payments for health care services were higher for Medicare beneficiaries aged 65 years and older with dementia than for other Medicare beneficiaries in the same age group (Table 1.2) [3].

Most patients with dementia live at home, usually with assistance from family and friends, but as their condition worsens, they receive increasing care from family or other unpaid caregivers [3]. Many of these patients

Age	Median survival in years from symptom onset (SD)	Median survival in years from initial diagnosis
<75 years	8.9 (4.5)	9.9
75–84 years	6.13 (3.3)	6.9
≥85 years	4.38 (2.46)	4.4

Table 1.1 Estimated survival time from symptom onset and time of diagnosis in patients with probable Alzheimer's disease. SD, standard deviation. Adapted from © American Medical Association, 2005. All rights reserved. Ganguli et al [6]. Adapted from © American Academy of Neurology, 2008. All rights reserved. Helzner et al [7]

	Beneficiaries with Alzheimer's disease or other dementia	Beneficiaries without Alzheimer's disease or other dementia
Inpatient hospital	$9,732	$3,912
Medical provider*	$5,967	$3,956
Skilled nursing facility	$3,812	$444
Nursing home	$17,693	$786
Hospice	$1,749	$171
Home health care	$1,402	$452
Prescription medications†	$2,681	$2,732

Table 1.2 Average per person payments for health care services provided to Medicare beneficiaries aged 65 and older with or without Alzheimer's disease or other dementia. From the 2008 Medicare Beneficiary Survey, in 2011 dollars. *'Medical provider' includes physician, other medical provider and laboratory services, and medical equipment and supplies. †Information on payments for prescription drugs is only available for people who were living in the community, that is, not in a nursing home or assisted living facility. Reproduced with permission from © the Alzheimer's Association, 2011. All rights reserved. The Alzheimer's Association [3].

also receive paid services at home or in adult day centers, assisted-living facilities, or nursing homes. The national average costs of these services are high: adult day center services, $70 per day; assisted-living facilities, $42,600 per year; and nursing home care, $90,520 (private room average cost) to $81,030 (semi-private room average cost) per year [11].

Definitions

The term dementia describes a condition in which there is a deterioration in the memory, thinking, behavior, and the ability of a person to perform everyday activities, which is not explained by a delirium or another medical disorder (Table 1.3) [12,13].

Although many different progressive neurodegenerative diseases can cause dementia, AD is the most common. For the first time since 1984, the definition of AD has been updated by an expert panel charged by the National Institute on Aging and the Alzheimer's Association [12,14,15]. These changes reflect the new scientific understanding that AD has a

Interfere with the ability to function at work or at usual activities
Represent a decline from previous levels of functioning and performing
Are not explained by delirium or major psychiatric disorder
The cognitive* or behavioral impairment involves a minimum of two of the following domains:
• Impaired ability to acquire and remember new information; symptoms include: repetitive questions or conversations, misplacing personal belongings, forgetting events or appointments, getting lost on a familiar route.
• Impaired reasoning and handling of complex tasks, poor judgment; symptoms include: poor understanding of safety risks, inability to manage finances, poor decision-making ability, inability to plan complex or sequential activities.
• Impaired visuospatial abilities; symptoms include: inability to recognize faces or common objects or to find objects in direct view despite good acuity, inability to operate simple implements, or orient clothing to the body.
• Impaired language functions (eg, speaking, reading, writing); symptoms include: difficulty thinking of common words while speaking, hesitations; speech, spelling, and writing errors.
• Changes in personality, behavior, or comportment; symptoms include: uncharacteristic mood fluctuations (such as agitation), impaired motivation, initiative, apathy, loss of drive, social withdrawal, decreased interest in previous activities, loss of empathy, compulsive or obsessive behaviors, socially unacceptable behaviors.

Table 1.3 Criteria for all-cause dementia. *Cognitive impairment is detected and diagnosed through a combination of (1) history taking from the patient and a knowledgeable informant and (2) an objective cognitive assessment, either a 'bedside' mental status examination or neuropsychological testing. Neuropsychological testing should be performed when the routine history and bedside mental status examination cannot provide a confident diagnosis. Reproduced with permission from © Elsevier, 2011. All rights reserved. McKhann et al [12].

lengthy pre-dementia prodrome, as well as new research information about the potential significance of biomarkers.

To encompass the full continuum of the disease as we now understand it, the panel developed guidelines for earlier stages of AD, mild cognitive impairment (MCI), 'preclinical' stages, and dementia due to AD. The dementia and MCI guidelines are offered for clinical use, whereas the preclinical guidelines were developed for research purposes and were intentionally made to be both provisional and flexible to allow for future advances from emerging technologies and understandings of biomarkers. Biomarkers are defined as physiologic, biochemical, or anatomic parameters that are measured in vivo that reflect specific features of disease-related pathophysiology. A comprehensive list of the major biomarkers under investigation in AD is provided in Table 1.4 [15].

The preclinical stage of Alzheimer's disease

The preclinical stage of AD describes the phase when clinical symptoms are not yet evident, but biologic markers of the disease are already

Biomarkers of Aβ deposition
CSF Aβ$_{42}$
PET amyloid imaging
Biomarkers of neuronal injury
CSF tau/phosphorylated-tau
Hippocampal volume or medial temporal atrophy by volumetric measures or visual rating
Rate of brain atrophy
FDG-PET imaging
SPECT perfusion imaging
Less well validated biomarkers: fMRI activation studies, resting BOLD functional connectivity, MRI perfusion, MR spectroscopy, diffusion tensor imaging, voxel-based and multivariate measures
Associated biochemical change
Inflammatory biomarkers (cytokines)
Oxidative stress (isoprostanes)
Other markers of synaptic damage and neurodegeneration, such as cell death

Table 1.4 Biomarkers under examination for Alzheimer's disease. Aβ, amyloid-beta protein; BOLD, blood oxygen level-dependent; CSF, cerebrospinal fluid; FDG, fluorodeoxyglucose; fMRI, functional magnetic resonance imaging; MR, magnetic resonance; MRI, magnetic resonance imaging; PET, positron emission tomography; SPECT, single photon emission tomography. Reproduced with permission from © Elsevier, 2011. All rights reserved. Albert et al [15].

present. Given the absence of clinical symptoms, biomarkers are necessary to establish the presence of the disease process, including amyloid-beta (Aβ) protein buildup and other early nerve cell changes. Because the biomarker data are not fully developed or standardized, the risk for progression to AD dementia is unknown at the preclinical stage. Thus, the use of imaging and biomarker tests at this stage are recommended only for research at this time [14].

Mild cognitive impairment due to Alzheimer's disease

The National Institute on Aging-Alzheimer's Association guidelines for MCI due to AD, although intended for research purposes, expand on existing clinical guidelines. The MCI stage is defined by symptoms of cognitive impairment (typically memory problems) that are evident and measurable but do not compromise independence as outlined in Table 1.5 [15]. To increase the likelihood that the underlying disease is a neurodegenerative disorder consistent with AD, it is necessary to rule out other systemic or brain diseases that could account for the decline in cognition, such as vascular, traumatic, or depressive conditions [15]. Not all people with MCI will progress to AD dementia, and thus the biomarkers can be used to establish the underlying etiology of clinical symptoms. Recommended biomarker tests include elevated levels of tau or decreased levels of Aβ in the cerebrospinal fluid (CSF), reduced glucose uptake in the brain as determined by fluorodeoxyglucose-positron

Establish clinical and cognitive criteria
Cognitive concern reflecting a change in cognition reported by patient, informant, or clinician (ie, historical or observed evidence of decline over time)
Objective evidence of impairment in one or more cognitive domains, typically including memory (ie, formal or bedside testing to establish level of cognitive function in multiple domains)
Preservation of independence in functional abilities
Not demented
Examine etiology of MCI consistent with AD pathophysiological process
Rule out vascular, traumatic, depressive causes of cognitive decline, where possible
Provide evidence of longitudinal decline in cognition, when feasible
Report history consistent with AD genetic factors, where relevant

Table 1.5 Summary of clinical and cognitive evaluation for mild cognitive impairment due to Alzheimer's disease. AD, Alzheimer's disease; MCI, mild cognitive impairment. Reproduced with permission from © Elsevier, 2011. All rights reserved. Albert et al [15].

emission tomography (FDG-PET), and atrophy of certain areas of the brain as measured with structural magnetic resonance imaging (MRI). Although intended primarily for the research setting, these tests may be applied in specialized clinical settings as a diagnostic adjunct to help determine possible causes of MCI symptoms [15].

Alzheimer's disease dementia

The 1984 clinical criteria for AD remains the foundation of the diagnosis in that progressive cognitive decline must be associated with impairment in functioning in daily activities. The new guidelines expand the concept of AD dementia beyond memory loss as its primary characteristic and incorporate the possibility of a decline in other aspects of cognition, such as anomia, visuospatial impairment, and impaired executive functioning, which may be the first symptoms to be noticed. Biomarkers may be used to enhance the specificity. AD dementia is classified into three categories [12]:

- probable AD dementia (Table 1.6);
- possible AD dementia (Table 1.7); and
- probable or possible AD dementia with evidence of supportive biomarkers

Insidious onset. Symptoms have a gradual onset over months to years, not sudden over hours or days

Clear-cut history of worsening of cognition by report or observation; and

The initial and most prominent cognitive deficits are evident upon history and examination in one of the following categories:

- Amnestic presentation: it is the most common syndromic presentation of AD dementia. The deficits should include impairment in learning and recall of recently learned information. There should also be evidence of cognitive dysfunction in at least one other cognitive domain, as defined earlier in the text.
- Nonamnestic presentations:
  – Language presentation: the most prominent deficits are in word-finding, but deficits in other cognitive domains should be present.
  – Visuospatial presentation: the most prominent deficits are in spatial cognition, including object agnosia, impaired face recognition, simultanagnosia, and alexia. Deficits in other cognitive domains should be present.
  – Executive dysfunction: the most prominent deficits are impaired reasoning, judgment, and problem solving. Deficits in other cognitive domains should be present.

Table 1.6 Criteria for probable Alzheimer's disease dementia. AD, Alzheimer's disease.
Reproduced with permission from © Elsevier, 2011. All rights reserved. McKhann et al [12].

Atypical course
Atypical course meets the core clinical criteria in terms of the nature of the cognitive deficits for AD dementia, but either has a sudden onset of cognitive impairment or demonstrates insufficient historical detail or objective cognitive documentation of progressive decline

Or

Etiologically mixed presentation
Etiologically mixed presentation meets all core clinical criteria for AD dementia but has evidence of: • concomitant cerebrovascular disease, defined by a history of stroke temporally related to the onset or worsening of cognitive impairment; or the presence of multiple or extensive infarctions or severe white matter hyperintensity burden; • features of dementia with Lewy bodies other than the dementia itself; or • evidence for another neurological disease or a nonneurological medical comorbidity or medication use that could have a substantial effect on cognition.

Table 1.7 Criteria for possible Alzheimer's disease dementia. AD, Alzheimer's disease.
Reproduced with permission from © Elsevier, 2011. All rights reserved. McKhann et al [12].

The first two categories are intended for clinical use, and the third is intended for research purposes at this time. Biomarkers are used to increase or decrease the level of certainty that AD is the cause of the dementia. As the understanding and validity of these tests improve, their application in clinical practice will increase.

References

1   Hebert LE, Scherr PA, Bienias JL, Bennett DA, Evans DA. Alzheimer disease in the US population: prevalence estimates using the 2000 census. *Arch Neurol.* 2003;60:1119-1122.
2   Nelson PT, Head E, Schmitt F, et al. Alzheimer's disease is not 'brain aging': neuropathological, genetic, and epidemiological human studies. *Acta Neuropathol.* 2011;12:571-587.
3   Alzheimer's Association. 2011 Alzheimer's Disease Facts and Figures. www.alz.org/downloads/facts_figures_2011.pdf. Accessed November 20, 2014.
4   World Health Organization. Dementia cases set to triple by 2050 but still largely ignored. http://www.who.int/mediacentre/news/releases/2012/dementia_20120411/en/. Accessed November 20, 2014.
5   Hebert LE, Beckett LA, Scherr PA, Evans DA. Annual incidence of Alzheimer disease in the United States projected to the years 2000 through 2050. *Alzheimer Dis Assoc Disord.* 2001;15:169-173.
6   Ganguli M, Dodge HH, Shen C, Pandav R, DeKosky ST. Alzheimer disease and mortality a 15-year epidemiological study. *Arch Neurol.* 2005;62:779-784.
7   Helzner EP, Scarmeas N, Cosentino S, Tang MX, Schupf N, Stern Y. Survival in Alzheimer's disease a multiethnic, population-based study of incident cases. *Neurology.* 2008;71:1489-1495.
8   Brookmeyer R, Corrada MM, Curriero FC, Kawas C. Survival following a diagnosis of Alzheimer disease. *Arch Neurol.* 2002;59:1764-1767.
9   Xie J, Brayne C, Matthews FE; Medical Research Council Cognitive Function and Ageing Study collaborators. Survival times in people with dementia: analysis from population based cohort study with 14 year follow-up. *BMJ.* 2008;336:258-262.

10 Arrighi HM, Neumann PJ, Lieberburg IM, Townsend RJ. Lethality of Alzheimer disease and its impact on nursing home placement. *Alzheimer Dis Assoc Disord*. 2010;24:90-95.

11 MetLife Mature Market Institute. Market Survey of Long-Term Care Costs: The 2012 MetLife Market Survey of Nursing Home, Assisted Living, Adult Day Services and Home Care Costs, October 2012. https://www.metlife.com/assets/cao/mmi/publications/studies/2012/studies/mmi-2012-market-survey-long-term-care-costs.pdf. Accessed November 20, 2014.

12 McKhann GM, Knopman DS, Chertkow H, et al. The diagnosis of dementia due to Alzheimer's disease: recommendations from the National Institute on Aging-Alzheimer's Association workgroups on diagnostic guidelines for Alzheimer's disease. *Alzheimers Dement*. 2011;7:263-269.

13 World Health Organization (WHO). Dementia Fact Sheet, No 362. WHO website. www.who.int/mediacentre/factsheets/fs362/en/. Accessed November 20, 2014.

14 Sperling RA, Aisen PS, Beckett LA, et al. Toward defining the preclinical stages of Alzheimer's disease: recommendations from the National Institute on Aging-Alzheimer's Association workgroups on diagnostic guidelines for Alzheimer's disease. *Alzheimers Dement*. 2011;7:280-292.

15 Albert MS, DeKosky ST, Dickson D, et al. The diagnosis of mild cognitive impairment due to Alzheimer's disease: recommendations from the National Institute on Aging-Alzheimer's Association workgroups on diagnostic guidelines for Alzheimer's disease. *Alzheimers Dement*. 2011;7:270-279.

Chapter 2

Diagnosing Alzheimer's disease

A diagnosis of Alzheimer's disease (AD) is a diagnosis of exclusion. A consideration of current and past medical and psychiatric conditions, surgeries, metabolic abnormalities, and pharmacotherapy is vital to making a diagnosis. Numerous factors can affect cognitive and functional abilities, and it is not unusual for other medical comorbidities to be present in addition to an underlying neurodegenerative process. Although these are rarely solely responsible for cognitive impairment, they may influence cognition adversely. This chapter will discuss recommended practices for the initial assessment of AD, including laboratory, imaging, and other types of diagnostic tests.

Initial evaluation

Numerous factors must be taken into account when evaluating a patient with possible dementia. It is critical to include the family in the discussion during an evaluation of someone with possible dementia, as a person with memory issues may not be aware of the degree of impairment they are suffering and, therefore, may not be able to accurately describe their level of function. Obtaining family members' input by interviewing them separately is critical to not only supplement the clinical history, but also to gain an understanding through candid discussions of the issues and concerns they have had or are experiencing. Education and support for the family begins with the first visit and remains a vital part of all subsequent visits to ensure continuity, an appropriate level of care, patient safety, and assess caregiver health and welfare.

© Springer Healthcare 2015
A. Burke et al., *Pocket Reference to Alzheimer's Disease Management*,
DOI 10.1007/978-1-910315-22-4_2

After a thorough medical assessment is completed and other medical conditions are excluded (Table 2.1), the clinician should look for patterns in the comprehensive history gathered from the patient, family, and caregivers. Vital information includes onset and type of symptoms as well as progression over time. Frequently, distinct patterns can be elicited from the patient's medical history in various dementias (Table 2.2). Additionally, the physician must take note of the various types of dementia that the patient may be experiencing (Table 2.3). A comprehensive evaluation of the patient history from both the patient and their family will supplement screening, laboratory, imaging, and other tests to determine which disorder the patient may have.

Medications affecting cognition	Medical conditions affecting cognition
Sleep aids	Head injury
Antihistamines	Stroke
Anticonvulsants	Sleep deprivation
Benzodiazepines	Substance abuse
Chemotherapeutics	Depression
Opiates	Anxiety
Antidepressants	Psychosis
Antihypertensives	Vitamin B_{12} deficiency

Table 2.1 Medications and medical conditions affecting cognition.

Cognitive changes	Difficulty with memory; difficulty finding words; disorientation; trouble understanding verbal or written communications; caregiver answers all questions for the patient during visits
Personality changes	Apathy; affective lability or blunting; social withdrawal; inappropriate friendliness; disinhibition; lower frustration tolerance; development of uncharacteristic personality traits; intensification of premorbid personality traits
Psychiatric symptoms	Suspiciousness; anxiety; depression; social withdrawal; hallucinations; delusions; irritability; odd beliefs; agitation; sundowning
Changes in function	Difficulty handling the finances; managing of calendar and appointments; shopping; using household appliances; driving; performing chores; or maintaining personal care/hygiene; getting lost

Table 2.2 Patterns associated with possible dementia. Adapted from © Lippincott Williams & Wilkins, 2013. All rights reserved. Burke et al [1].

Disorder	Onset	Progression	Age of onset	Most prominent symptoms	Frequency
Alzheimer's dementia	Insidious	Gradual decline	>60 years	Initially begins with short-term memory impairment	≈60–80% of all dementias
Vascular dementia	Abrupt	Stepwise decline	>60 years	Memory and behaviors may be affected depending on number and location of lesions	≈10–20% of all dementias
Frontotemporal dementia	Insidious	Gradual decline	35–75 years	Limited memory impairment initially (<10%) **Behavioral variant:** prominent impairments in judgment, insight, and impulse control; executive dysfunction; and socially inappropriate behaviors **Primary progressive aphasia:** word finding difficulty and effortful speech with intact language comprehension **Semantic dementia:** difficulty understanding words and naming or difficulty recognizing faces and objects	<10% of all dementias
Dementia with Lewy bodies	Insidious	Gradual decline	Mean: 75 years	Short-term memory impairment and impairment in visuospatial abilities; parkinsonian symptoms occur within first year of cognitive impairment; frequent fluctuations of cognition/alertness; vivid visual hallucinations of mute animate objects; REM sleep disturbances; sensitivity to antipsychotics	≈10–22% of all dementias
Parkinson's disease dementia	Insidious	Gradual decline	Mean: 75 years	Symptoms are similar to dementia with Lewy bodies but dementia occurs years after onset of parkinsonian symptoms	Affects ≈20% of patients with Parkinson's disease

Table 2.3 Differential patterns in dementia. REM, rapid eye movement. Adapted from © American Medical Association, 1997. All rights reserved. Small et al [2]. Adapted from © American Psychiatric Association, 1997. All rights reserved. American Psychiatric Association [3]. Adapted from © Elsevier, 1994. All rights reserved. Morris [4].

Confusion: delirium, dementia, or both?

As noted above, when a physician is faced with a patient with possible AD, many different mental disorders, such as delirium and dementia, must be taken into account as several symptoms of these disorders overlap and are comorbid. Delirium is a common neuropsychiatric syndrome in the elderly, characterized by concurrent impairment in cognition and behavior. Delirium is an acute change in mental status, developing over hours to days; Table 2.4 further describes its symptoms. Etiologies of delirium are often multifactorial and are due to medications (eg, anticholinergics, steroids, anti-epileptics, lanoxin, benzodiazepines, opioids, sedatives/hypnotics, dopamine and dopamine agents, alcohol, and recreational drugs), pain, fecal impaction, hypoxia, renal failure, urinary retention, urinary tract infection, pneumonia and other infections, myocardial infection, acute cerebrovascular accident, metabolic changes (eg, $\uparrow$Ca++, $\downarrow$Na, elevated or decreased glucose levels), and environmental changes.

Delirium is often missed or mislabelled as dementia (Table 2.5) or depression. Untreated delirium may have severe consequences in the elderly, with high rates of morbidity and mortality. Evidence indicates that

Symptoms
Clouding of state of consciousness
Disorganized thinking
Disorientation to time/place
Inattention or distraction
Drowsiness, disturbed sleep
Increase or decrease in psychomotor activity
Fluctuation of signs and symptoms
Improvement/normalization of mental status after treatment of underlying condition
Frequency of delirium
Occurs in about half of elderly patients during or after acute hospitalization; more than half of these patients are found to have permanent, previously undiagnosed dementia
Up to 70% of intensive care patients develop delirium
Delirium occurs in 28–83% of patients near the end of life

Table 2.4 Symptoms and frequency of delirium. Terminal agitation or restlessness are other terms used to describe delirium and severe agitation in the last days of life. Reproduced with permission from © Elsevier, 2008. All rights reserved. Maldonado [5].

Delirium	Dementia
Abrupt onset: hours to days	Insidious onset: months to years
Duration: days to weeks	Duration: 2–20 years
Symptoms include confusion, forgetfulness, altered sleep–wake cycle, high frequency of delusions, hallucinations, and illusions	Symptoms include decline in memory, speech and language difficulty, loss of self-care abilities, and a variety of behavioral problems; causes depend on etiology of disease
Causes include acute medical conditions, medications, alcohol abuse, acute psychosis	

Table 2.5 Differentiating dementia and delirium. Reproduced with permission from © Sage Publications, 2011. All rights reserved. Mittal et al [6].

early detection, reduction of risk factors, and better patient management can decrease morbidity rates [6].

To determine which medical condition the patient has, it is essential to assess their mental status. Formal methods may be used, such as the Mini-Mental Status Exam (MMSE), Short Portable Mental Status Questionnaire (SPMSQ), Clock-Drawing Test, and Confusion Assessment Method, to assess for areas of strength and weakness in the thought process. Attention difficulties are prominent in delirium and may in turn affect other aspects of thought. Additionally, the patient may be asked a series of questions or prompts to determine their level of delirium and/or dementia; for example, asking the patient:
- their orientation to name, date, place (city, state);
- to repeat three items (eg, hat, tree, car);
- to spell 'world' backward; and
- to repeat three items again.

The physician can then evaluate their patient's response; for example, was there a change in mental status from baseline? Does the behavior fluctuate? Can the patient focus his/her attention? Is the patient's thinking disorganized or incoherent? Are they switching thoughts/ideas? Can the patient follow commands?

It is also critical to assess the patient's behaviors by having the patient and caregivers describe the actual behaviors or symptoms demonstrated and the frequency of which they occur. It is equally important to establish if the behaviors are new or have increased in frequency over time. The physician must also obtain from family and caregivers any successful interventions that they have attempted to use for the identified behaviors (see Chapter 5). Common behaviors associated with delirium and dementia are described in Table 2.6.

Staging Alzheimer's-type dementias

For primary care providers, a four-point trajectory is suggested for staging AD. The trajectory begins with the presentation of mild cognitive impairment (MCI). This is followed by the three stages of dementia: mild, moderate, and severe. Most staging guides have been developed for people with AD dementia; those with non-AD dementias may not fit into these stages.

Staging dementia depends on triangulating several data sources:
- the family report of function and behaviors;
- the scores from a standardized cognitive screening instrument, such as the MMSE or Montreal Cognitive Assessment (MoCA) tools; and
- interviewing the person.

For an accurate diagnosis of stage, all three data sources must be considered. Level of function is very consistent in determining disease trajectory

Delirium	Dementia
Hyperactive delirium	**Moderate dementia**
Hallucinations, delusions, paranoia	Resisting/fighting hands-on caregivers
Agitation, fidgeting	Assaultive toward caregivers/peers
Pulling at clothing, dressings, tubes	Wandering and rummaging
Crawling out of bed	Physical restlessness
	Sundowning
	Eating problems
	Sleeping problems
	Yelling
	Sexual behaviors
Hypoactive delirium	**Advanced dementia**
Difficult to arouse, stuporous	Resisting/fighting hands-on caregivers
	Fall risk (wanting to walk when unable to)
	Physical restlessness
	Resisting/refusing to eat/drink
	Disruptive sleep patterns
	Disruptive yelling

Table 2.6 Common behaviors in delirium and dementia. Reproduced with permission from © Elsevier, 2008. All rights reserved. Maldonado [5].

across persons. One may infer specific noncognitive symptoms with losses of each activity. Symptoms of MCI include [7]:

- no functional decline (thus no dementia is present);
- complaints of memory changes;
- possible depression or apathy;
- frustration and irritability – the person may develop a 'short fuse,' becoming angry more easily and having new conflicts with others (especially marital conflicts);
- slight declines in short-term memory, with a decreased ability to learn new things and retain new information;
- increased self-centered or self-absorbed behavior (interpersonal relationships, especially marital relationships may change as a result);
- an increase in intensity of emotions, including sadness or happiness over seemingly minor things;
- development of problems with employer or poorer job performance;
- decreased sense of time, characterized by obsessing over appointment times, getting dressed early for scheduled events, and worrying about when things will occur; and
- MMSE score of 30–26 (MMSE scoring will be further discussed on pages 24–25).

The following figures illustrate the degenerative changes in the brain associated with each of the three stages: early/pre-dementia (MCI), years 1–2 of disease (Figure 2.1A); mild-to-moderate changes, years 3–6 (Figure 2.1B); and advanced changes, year 7–death (Figure 2.1C) [8].

Neuropsychological testing
Dementia and neuropsychology
In order to better isolate which of the patient's abilities may have been compromised or affected, a series of tests known as a 'neuropsychological battery' may be administered (most likely by a referred neuropsychologist). A typical neuropsychological evaluation might focus on measuring various abilities, such as: general intelligence; attention and concentration; learning and memory; motor and sensory functioning; auditory and visual processing; language functions; thinking; planning

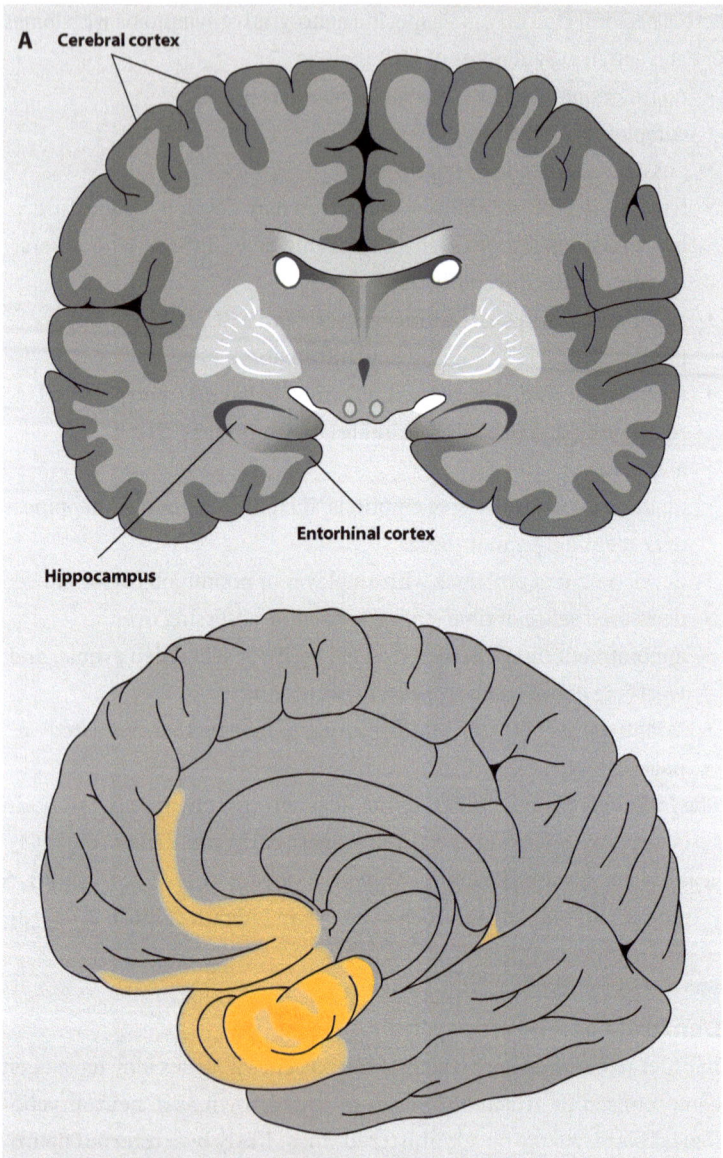

Figure 2.1A Brain deterioration in early Alzheimer's disease dementia, years 1–2.
Shading indicated affected areas. Adapted from © Sage Publications, 2011. All rights reserved.
Mittal et al [6]. Adapted from © National Institute of Aging, National Institutes of Health, 2011.
All rights reserved. National Institute of Aging, National Institutes of Health [8].

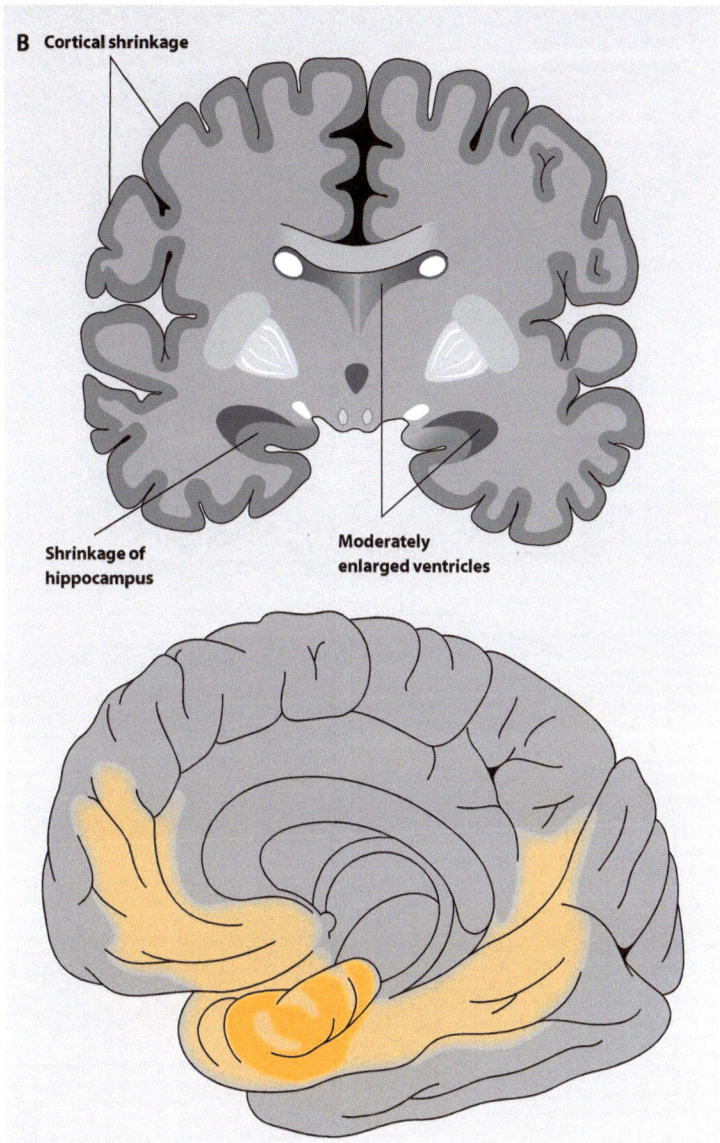

B Cortical shrinkage

Shrinkage of hippocampus

Moderately enlarged ventricles

Figure 2.1B Brain deterioration in mild-to-moderate Alzheimer's disease dementia, years 3–6. Shading indicated affected areas. Adapted from © Sage Publications, 2011. All rights reserved. Mittal et al [6]. Adapted from © National Institute of Aging, National Institutes of Health, 2011. All rights reserved. National Institute of Aging, National Institutes of Health [8].

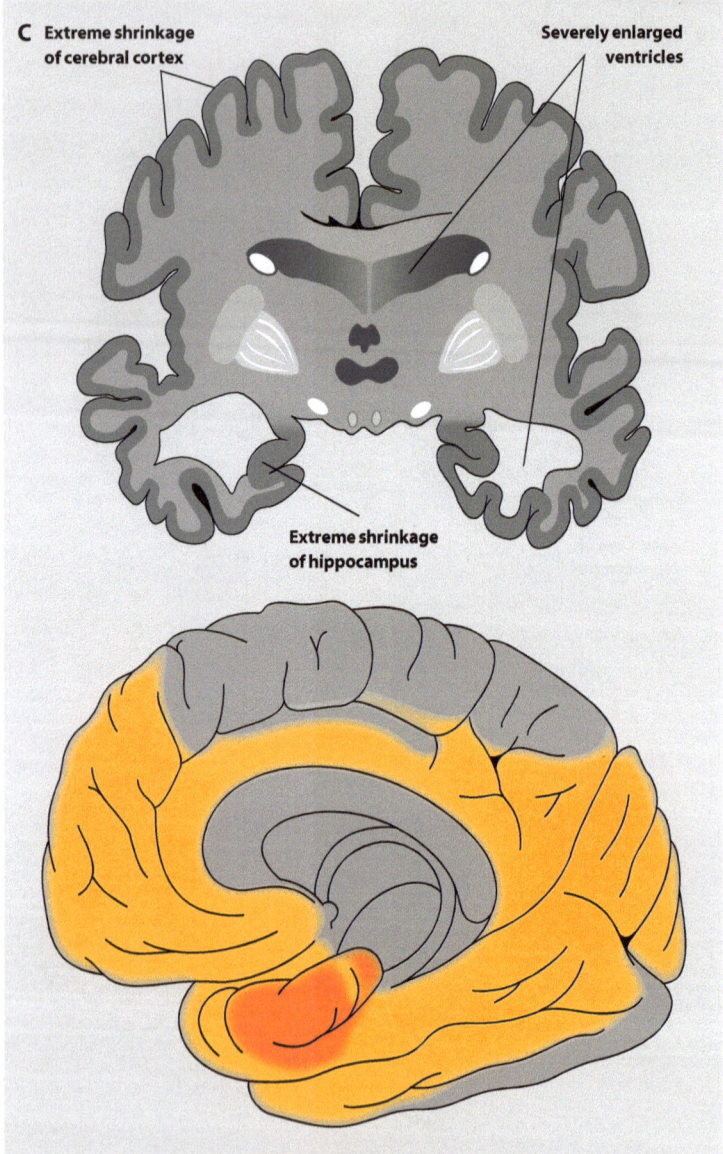

Figure 2.1C Brain deterioration in advanced Alzheimer's disease dementia. Shading indicated affected areas. Adapted from © Sage Publications, 2011. All rights reserved. Mittal et al [6]. Adapted from © National Institute of Aging, National Institutes of Health, 2011. All rights reserved. National Institute of Aging, National Institutes of Health [8].

and organization; speed of processing; executing functioning; expressive functions; and emotions and personality [9].

What to expect when administering a neuropsychological test battery

In general, an elderly patient may expect to spend approximately 2–3 hours taking a neuropsychological test battery, plus additional time may be needed to interview the patient or caregivers. In some cases, extra testing may be required, which would lengthen the amount of testing time. Breaks are given upon request or as needed. Tests are generally short in nature, taking 5–10 minutes each and covering areas related to intelligence, attention and concentration, learning and memory, language abilities, and organizational skills among others.

When neuropsychological batteries are not recommended

There are many reasons why a neuropsychological test battery may be inappropriate or not recommended. The list below highlights some of the reasons why a test battery might be unnecessary:

- very low scores on cognitive screening tests;
- acute psychosis or delirium; and/or
- lack of fluency in the language of the neuropsychologist.

Additionally, neuropsychologic batteries could frustrate the patient without reason and may not provide any other useful information. Plus, the accommodations needed to perform the battery could invalidate the testing norms and prove unreliable [10].

Screening tests

Screening tests are quick and convenient assessments that measure a restricted set of cognitive abilities, such as, but not limited to, memory, language, mental clarity, and attention. Instructions and scoring procedures are easily learned, which adds to their appeal. Useful screening tests and their strengths and weaknesses in regards to detection of dementia and MCI are listed in Table 2.7 [9,11]. Table 2.8 has a classification of each stage of AD by MMSE score.

The need for more extensive evaluation

While extremely useful, screening tools have limitations and can provide scores that suggest impairment, but are not conclusive for dementia. A more thorough examination is needed whenever there is doubt about a patient's cognitive functioning. However, when dealing with a geriatric population, it is important to keep neuropsychological testing short and comprehensive. An important balance needs to be achieved between keeping the patient motivated and involved, while also collecting essential data; this helps to ensure that the collected data are accurate representations of the patient's status without fatigue or disinterest becoming a factor. A neuropsychological evaluation can achieve the following:

- provide a baseline for future testing;
- aid in the differentiation of the various dementia presentations;
- identify compensatory strategies;
- assist in judging if deficits are organic or psychiatric in nature;
- aid in earlier detection of 'preclinical' dementia, such as MCI;
- describe patterns of cognitive weaknesses and strengths; and
- assist in choosing treatments and preventative/postponing measures.

In isolation, neuropsychological tests generally cannot differentiate between various dementia conditions, but together with historical data, clinical

Exam	Strengths	Weaknesses
Mini-Mental Status Exam (MMSE)	Able to identify those with moderate-to-severe deficits; less sensitive in identifying mild dementia and MCI; is the most widely used screening test	May misidentify patients ≥60 years old with <9 years of education as demented or patients with higher verbal intelligence as cognitively impaired
Montreal Cognitive Assessment (MoCA)	More sensitive than MMSE to patients with mild dementia or MCI	Conclusions regarding validity are restricted to memory clinic studies
Rey Auditory Verbal Learning Test (RAVLT)*	Patients with early AD have very low recall on trial I; recall ≈6 words by trial V; and have difficulty remembering words after distraction	Measures only one specific cognitive domain-episodic declarative memory

Table 2.7 Useful screening tests for detecting dementia and mild cognitive impairment.
*The RAVLT is a five-trial presentation of a 15-word list that provides an analysis of learning and retention followed by a distraction list. In studies, patients with AD have exhibited consistent learning and retention profiles. AD, Alzheimer's disease; MCI, mild cognitive impairment.
Adapted from © Oxford University Press, 2012. All rights reserved. Lezak et al [9].

data, and observations they can help by providing insight into specific cognitive abilities that are markers of certain dementias and other disorders. Table 2.9 shows types of dementia, clinically-related disorders, and possible characteristics of differentiation [12].

Laboratory tests
Routine laboratory tests for Alzheimer's disease
Because no laboratory test currently exists that can diagnose AD while the patient is alive, laboratory testing is used to exclude alternate etiologies of dementia or medical comorbidities that can exacerbate symptoms [13]. Relevant reversible causes of dementia overlap with causes of delirium, such as infection, metabolic or endocrine abnormalities, or side effects of medications. Blood tests that are generally recommended for routine work-up in patients with dementia are presented in Table 2.10 [13].

Nonroutine laboratory tests for Alzheimer's disease
Nonroutine laboratory tests that may be useful in certain circumstances include the following:
- erythrocyte sedimentation rate;
- urinalysis;
- toxicology;
- HIV testing;
- syphilis serology;
- cerebrospinal fluid (CSF) examination;
- electroencephalogram; and
- fluorodeoxyglucose-positron emission tomography (FDG-PET).

Image testing
Routine imaging for Alzheimer's disease
The routine work-up for AD includes structural brain imaging with non-contrast computed tomography (CT) or preferably magnetic resonance imaging (MRI) of the brain [13]. The utility of structural brain imaging is limited to the exclusion of other etiologies, such as tumors, subdural hematomas, strokes, or hydrocephalus. Common features of AD seen in structural imaging are cortical atrophy and ventricular enlargement

Mild MMSE score of 26–20	Moderate MMSE score of 19–10	Severe MMSE score of 9–0
Memory and thinking		
• Difficulty with short-term memory • Poor concentration • Poor decision-making • Problems with understanding time • Loss of executive function • Inability to function appropriately in emergencies • Loss of ability to live alone	• Difficulty with short- and long-term memory • Forgets own personal history • May begin to forget friends and family members • No sense of time • Repetitive questions and behaviors	• Severely impaired memory for recent and past events • Has periods of relative lucidity • Unable to follow simple commands
Language		
• Problems remembering the right word or name • Diminished reading comprehension	• May not understand what is being said • Losing ability to express themselves and making needs known • Little or no reading comprehension	• Unable to carry on a meaningful conversation • May cry out spontaneously
Mood		
• May become depressed or socially withdrawn	• More easily upset and frustrated • May appear to lack emotion • Fatigues quickly (<90 minutes) in social or high-stimulus activities and demands to leave • Onset of late-day confusion and agitation • Needs increased daytime rests to maintain function and mood • Loss of recognition of family and home (late in this stage)	• Appears withdrawn • Difficult to engage • Little eye contact

Function

- Trouble handling finances
- Difficulty initiating activities
- Gets lost/mixed-up when driving in familiar places
- May be involved in 'fender benders'
- Changes in socialization
- Limited in doing instrumental activities in their usual order:
  - Money management
  - Employment
  - Driving
  - Shopping
  - Medication management and administration
  - Home maintenance
  - Meal preparation
  - Running the thermostat

Needs help doing the following activities in their usual order:
- Bathing:
  - Forgets to bathe
  - May become afraid of water in bath/shower
  - May become resistant to bathing
- Grooming:
  - Lack of attention to fine details
  - Needs reminders for shaving, tooth brushing
  - Unable to apply make-up
- Selecting clothing:
  - Wears same clothing day after day
  - Changes clothing often in a day (confused by too many choices in closet)
  - Sleeps in clothing
- Dressing:
  - Selecting/coordinating clothing
  - Sequencing clothing
  - Buttoning, zipping, snapping clothing
- Bladder and bowel:
  - Difficulty finding the toilet
  - Does not recognize mirror image, thus evacuates in inappropriate places
  - Forgets to wipe and/or flush
  - Incontinent episodes
- Ambulating:
  - Shuffles
  - Postural change (eg, leans forward)
  - Frequent falls
- Eating:
  - Forgets to eat or drink
  - Forgets how to use silverware
  - May lack table manners

- Has difficulty interacting/responding to surroundings
- Forgets how to walk without help; may lead to eventual loss of body movement
- Relies totally on caregivers for:
  - Dressing
  - Grooming
  - Bathing
  - Feeding
  - Bladder/bowel etiquette
- May forget to chew food or swallow
- May lose ability to sit up, hold head up, and/or smile

Table 2.8 Function and symptoms by Mini-Mental Status Exam stages. MMSE, Mini-Mental Status Exam.

Dementia or disorder	Characteristics of differentiation
AD dementia	Particularly impaired on measures of delayed recall; produce more intrusion errors after interference; show consistent loss of semantic knowledge
	Deficits in episodic memory that are significantly greater than their executive functioning deficits
Vascular dementia	Patients with subcortical vascular dementia are more impaired on tests of executive functions and less impaired on episodic memory than AD dementia patients (particularly delayed recall)
Dementia with Lewy bodies	Prominent deficits in visuoperceptual and visuoconstructive abilities compared to patients with AD dementia
Frontotemporal dementia (eg, dementia due to Picks disease)	Patients with frontotemporal dementia performed significantly worse than patients with AD dementia on word producing tests, such as letter and category fluency tests
	Letter fluency performance was worse than semantic fluency performance in patients with frontotemporal dementia (the opposite was true for patients with AD dementia)
	Patients with frontotemporal dementia performed better on memory tests and visuospatial abilities (eg, block design and clock drawing) than patients with AD dementia
Dementia due to Huntington's disease	Patients with Huntington's disease are less likely to be impaired on recognition memory testing than patients with AD dementia
Delirium	Delirium is marked by disturbance in consciousness not usually seen in dementia patients
	Tests that target attention, such as Digit Span, Trail Making Test A, and cancellation tasks, would be helpful in differentiating it from dementia

Table 2.9 Dementias, disorders, and possible characteristics of differentiation. AD, Alzheimer's disease. Adapted from ©Atypon Literatum, 2009. All rights reserved. Salmon, Bondi [12]. Adapted from ©American Psychiatric Publishing, 2009. All rights reserved. Fearing, Inouye [14].

Test	Purpose
Complete blood count	To rule out anemia, infection
Complete metabolic panel	To rule out renal or hepatic dysfunction, electrolyte abnormalities, abnormal glucose
Vitamin B12 level	To rule out vitamin B12 deficiency
Thyroid function test	To rule out thyroid disease

Table 2.10 Blood tests for a routine dementia work-up. Adapted from ©American Academy of Neurology, 2001. All rights reserved. Knopman et al [13]

(Figure 2.2), which are neither sensitive nor specific markers of AD and can often accompany normal aging [15].

Fluorodeoxyglucose-positron emission tomography

FDG-PET imaging shows bilateral temporoparietal hypoperfusion (Figure 2.3) [15]. Current clinical dementia guidelines do not recommend routine FDG-PET scans in dementia evaluations [13]. However, it is useful for distinguishing between clinically ambiguous cases of AD versus frontotemporal dementia. The European Federation of the Neurological Societies (EFNS) recommends the use of FDG-PET imaging as part of the routine diagnostic work-up of clinically questionable dementia cases [16]. In vivo imaging of amyloid plaques using PET scans with radiolabeled ligands specific to amyloids is a recently approved diagnostic tool. Although a positive amyloid scan does not denote a diagnosis of AD, it indicates a greater likelihood of the disease.

Other testing

Lumbar puncture

Cerebrospinal fluid (CSF) examination can be useful in the appropriate clinical setting to rule out infection or other disease, such as multiple sclerosis. Although used routinely in the research setting, CSF biomarker analysis can be used as an adjunct test in the clinic in diagnostically challenging cases. Amyloid beta (Aβ) levels in the CSF, in particular $A\beta_{1-42}$

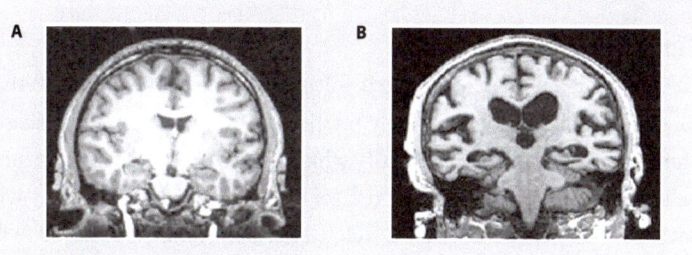

Figure 2.2 Coronal magnetic resonance images through the temporal lobes.
A, A normal subject; **B,** a patient with AD. The cerebral cortex and hippocampi of the patient with AD are visibly atrophic with ventricular enlargement compared with the normal subject. AD, Alzheimer's disease. Reproduced with permission from © Thieme Medical Publishers. year. All rights reserved. Yaari, Corey-Bloom [15].

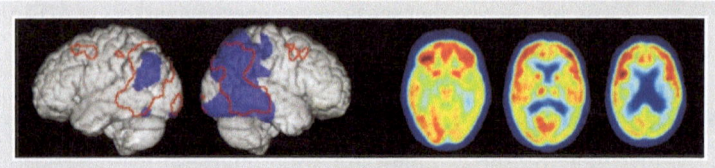

Figure 2.3 Fluorodeoxyglucose-positron emission tomography imaging shows bilateral temporoparietal hypoperfusion. Fluorodeoxyglucose-positron emission tomography scans of an individual with dementia due to AD. Images on the left are individual FDG-PET CMRgl binding showing areas of significant glucose hypometabolism compared to normal controls (blue). An automated algorithm was used to transform individual patient images into the dimensions of a standard brain and compute statistical maps of significantly reduced glucose metabolism relative to 67 normal control subjects (mean age 64). Red outlined regions represent areas of mean hypometabolism seen in FDG-PET scans from 14 patients with AD (mean age 64), compared to the same 67 normal controls. On the right are raw FDG-PET color maps from the same patient. Here we can see the use of FDG-PET for identifying disease-specific patterns of glucose metabolism for clinical use in individual patients to assist with diagnostic decision-making. AD, Alzheimer's disease; CMRgl, cerebral metabolic rate of glucose metabolism; FDG-PET, fluorodeoxyglucose-positron emission tomography. Adapted from © Thieme Medical Publishers. year. All rights reserved. Yaari, Corey-Bloom [15].

(Aβ42), are reduced by 40–50% in patients with AD compared to normal controls [17]. CSF total tau (T-tau) has a two- to three-fold elevation in patients with AD, which has been demonstrated in cross-sectional studies [18]. A ratio between T-tau to Aβ42 levels has high sensitivity (89%) and specificity (90%) to determine if a patient has AD [17].

The great majority of AD cases are detected as late-onset in patients with no known causal gene. The *apolipoprotein Eε4 (ApoE-ε4)* allele is a genetic susceptibility gene that increases the risk of AD. Heterozygotes of this allele have a 3-times greater risk of developing AD; homozygotes have a 15-times greater risk [19].

Genetic testing

There is such a variety of factors involved in the development of AD that genetic testing cannot accurately predict who will develop the disease. Thus, while blood tests can identify which *ApoE* alleles a person has, and indicate who may be at a higher risk for developing the disease, they will not show who will definitively develop AD. ApoE testing is believed to be more useful for studying risk for AD in large groups, not for determining one person's risk factor. In clinicians' offices and other clinical settings, genetic testing is used primarily for people with a family history of early-onset AD. It is not generally recommended for people at risk of late-onset AD.

Early-onset Alzheimer's disease

Early-onset familial dementias are rare and include mutations concerning the Aβ precursor protein (APP) and *presenilin 1* and *2* (*PS-1* and *PS-2*) genes [20]. Most cases are inherited as a type known as familial AD (FAD), and are caused by single gene mutations. These mutations cause abnormal proteins to form, ultimately contributing to the harmful amyloid plaques which are distinguishing characteristics of the disease. A child whose mother or father carries a genetic mutation for FAD has a 50/50 chance of inheriting that mutation. If the mutation is in fact inherited, the child almost surely will develop FAD. Predictive testing for family members of affected individuals is possible. It is a personal choice, which should be done in a setting that allows for informed consent, genetic and psychological counseling, and confidentiality.

Late-onset Alzheimer's disease

Most cases of AD are considered late-onset forms of the disease and occur after the age of 60. The causes of late-onset AD likely include a combination of genetic, environmental, and lifestyle factors that contribute to a person's risk for developing the disease [21].

The single gene mutations associated with early-onset AD do not seem to be involved in late-onset AD. A specific gene that causes late-onset AD has not been identified, but one genetic risk factor that does appear to correlate with a person's risk of developing the disease is the *ApoE* gene found on chromosome 19, which comes in several different forms or alleles. The more common forms are:

- *ApoE-ε2*: relatively rare; may provide some protection against the disease. Alzheimer's disease would more likely develop later in life for a person with this allele rather than if the person had other *ApoE* genes;
- *ApoE-ε3*: the most common; is thought to neither decrease nor increase the risk for the development of AD; and
- *ApoE-ε4*: occurs in 25–30% of the population and approximately 40% of all people with late-onset AD have this gene.

However, *ApoE-ε4* is only a genetic risk factor which, for still unknown reasons, increases the risk of developing AD. People who inherit one or two *ApoE-ε4* alleles tend to develop the disease at an earlier age than those who do not have any *ApoE-ε4* alleles. However, inheriting an *ApoE-ε4* allele does not mean that a person will definitely develop AD. Some people with one or two *ApoE-ε4* alleles never get the disease, and others who develop AD do not have any *ApoE-ε4* alleles.

References

1 Burke A, Hall G, Tariot PN. The clinical problem of neuropsychiatric signs and symptoms in dementia. *Continuum (Minneap Minn)*. 2013;19(2 Dementia):382-396.

2 Small GW, Rabins PV, Barry PP, et al. Diagnosis and treatment of Alzheimer disease and related disorders. Consensus statement of the American Association for Geriatric Psychiatry, the Alzheimer's Association, and the American Geriatrics Society. *JAMA*. 1997;278(16):1363-1371.

3 American Psychiatric Association. Practice guideline for the treatment of patients with Alzheimer's disease and other dementias of late life. American Psychiatric Association. *Am J Psychiatry*. 1997;154(5 suppl):1-39.

4 Morris JC. Differential diagnosis of Alzheimer's disease. *Clin Geriatr Med*. 1994;10:257-276.

5 Maldonado JR Delirium in the acute care setting: characteristics, diagnosis and treatment. *Crit Care Clin*. 2008;24:657-722

6 Mittal V, Muralee S, Williamson D, et al. Review: delirium in the elderly: a comprehensive review. *Am J Alzheimers Dis Other Demen*. 2011;26:97-109.

7 Gauthier S, Reisberg B, Zaudig M, et al. Mild cognitive impairment. *Lancet* . 2006;367: 1262-1270.

8 National Institute on Aging, National Institutes of Health. Alzheimer's disease: unraveling the mystery. NIH publication 08-3782. Published January 2011. www.nia.nih.gov/sites/default/ files/alzheimers_disease_unraveling_the_mystery.pdf. Accessed November 20, 2014.

9 Lezak MD, Howieson, DB, Loring DW, eds. *Neuropsychological Assessment*. 5th edition. New York, NY: Oxford University Press; 2012.

10 Hill-Briggs F, Dial JG, Morere DA, Joyce A. Neuropsychological assessment of persons with physical disability, visual impairment or blindness, and hearing impairment or deafness. *Arch Clin Neuropsychol*. 2007;22:389-404.

11 Strauss E, Sherman EMS, Spreen O. *A Compendium of Neuropsychological Tests: Administration, Norms, and Commentary*. 3rd edition. New York, NY: Oxford University Press: 2006.

12 Salmon DP, Bondi MW. Neuropsychological assessment of dementia. *Annu Rev Psychol*. 2009;60:257-282.

13 Knopman DS, DeKosky ST, Cummings JL, et al. Practice parameter: diagnosis of dementia (an evidence-based review). Report of the Quality Standards Subcommittee of the American Academy of Neurology. *Neurology*. 2001;56:1143-1153.

14 Fearing MA, Inouye SK. Delirium. In: Blazer DG, Steffens DC, eds. *The American Psychiatric Publishing Textbook of Geriatric Psychiatry*. Washington, DC: American Psychiatric Publishing; 2009:229-241.

15 Yaari R, Corey-Bloom J. Alzheimer's disease. *Semin Neurol*. 2007;27:32-41.

16 Hort J, O'Brien JT, Gainotti G, et al; EFNS Scientist Panel on Dementia. EFNS guidelines for the diagnosis and management of Alzheimer's disease. *Eur J Neurol*. 2010;17:1236-1248.

17 Thal LJ, Kantarci K, Reiman EM, et al. The role of biomarkers in clinical trials for Alzheimer disease. *Alzheimer Dis Assoc Disord*. 2006;20:6-15.

18 Hampel H, Buerger K, Zinkowski R, et al. Measurement of phosphorylated tau epitopes in the differential diagnosis of Alzheimer disease: a comparative cerebrospinal fluid study. *Arch Gen Psychiatry*. 2004;61:95-102.

19 Farrer LA, Cupples LA, Haines JL, et al. Effects of age, sex, and ethnicity on the association between apolipoprotein E genotype and Alzheimer disease: a meta-analysis. *JAMA*. 1997;278:1349-1356.

20 Cruchaga C, Chakraverty S, Mayo K, et al; for the NIA-LOAD/NCRAD Family Study Consortium. Rare variants in APP, PSEN1 and PSEN2 increase risk for AD in late-onset Alzheimer's disease families. *PLoS One*. 2012;7:e31039.

21 Wang S-C, Oelze B, Schumacher A. Age-specific epigenetic drift in late-onset Alzheimer's disease. *PLoS One*. 2008;3:e2698.

Chapter 3

Pharmacological treatment of cognitive decline in Alzheimer's disease

Treatment goals and currently available therapies

A primary target of pharmacologic therapy for Alzheimer's disease (AD) is the cognitive symptoms associated with the condition, including dementia. Cognitive treatment goals may include:

- improving memory;
- improving functional status;
- improving behavioral symptom;
- slowing progression; and/or
- delaying or preventing symptomatic onset.

Currently, there are four pharmacologic agents approved in the US for the treatment of dementia associated with AD (Table 3.1) [1,2]. Donepezil, galantamine, and rivastigmine are cholinesterase inhibitors, and memantine is an *N*-methyl-D-aspartate (NMDA) receptor antagonist. Tacrine was the first approved cholinesterase inhibitor; however, it has been discontinued in the US and is rarely prescribed due to safety concerns [1]. No drugs are currently approved for the treatment of mild cognitive impairment (MCI) associated with AD [3].

Cholinesterase inhibitors

Acetylcholine is predominantly found in the basal forebrain and memory networks of the brain. Loss of cholinergic function and activity is a hallmark of AD, with greater declines occurring as the disease gets more

© Springer Healthcare 2015
A. Burke et al., *Pocket Reference to Alzheimer's Disease Management*,
DOI 10.1007/978-1-910315-22-4_3

Drug	Donepezil	Galantamine	Rivastigmine	Memantine
Mechanism of action	Cholinesterase inhibitors	Cholinesterase inhibitors	Cholinesterase inhibitors	NMDA-receptor antagonist
FDA-approved indication	Mild-to-moderate AD; moderate-to-severe AD	Mild-to-moderate AD	Mild-to-moderate AD	Moderate-to-severe AD
Generic form available	Orally disintegrating tablet	Tablet, oral solution, ER capsule	Capsule only	NA
Recommended dosage	Orally disintegrating tablet: initial dose of 5 mg/day; may increase dose to 10 mg/day after 4–6 weeks if well tolerated, then to 23 mg/day after at least 3 months 23-mg dose available as brand-name tablet only	Tablet, oral solution: initial dose of 8 mg/day (4 mg twice a day); may increase dose to 16 mg/day (8 mg twice a day) and 24 mg/day (12 mg twice a day) at minimum 4-week intervals if well tolerated ER capsule: same dosage as above but taken once a day	Capsule, oral solution: initial dose of 3 mg/day (1.5 mg twice a day); may increase dose to 6 mg/day (3 mg twice a day), 9 mg (4.5 mg twice a day), and 12 mg/day (6 mg twice a day) at minimum 2-week intervals if well tolerated Patch: initial dose of 4.6 mg/day; may increase to 9.5 mg once a day and 13.3 mg once a day at minimum 4-week intervals if well tolerated	Tablet/oral solution: initial dose of 5 mg/day; may increase dose to 10 mg/day (5 mg/twice daily), 15 mg/day (5 mg and 10 mg as separate doses), and 20 mg/day (10 mg twice a day) at minimum 1-week intervals if well tolerated ER tablet: initial dose of 7 mg once a day; may increase dose to 14 mg/day, 21 mg/day, and 28 mg/day at minimum 1-week intervals if well tolerated
Common side effects	Nausea, vomiting, diarrhea	Nausea, vomiting, diarrhea, weight loss, loss of appetite	Nausea, vomiting, diarrhea, weight loss, loss of appetite, muscle weakness	Dizziness, headache, constipation, confusion

Table 3.1 Current therapies for dementia symptoms of Alzheimer's disease. AD, Alzheimer's disease; ER, extended release; FDA, US Food and Drug Administration; NA, not available; NMDA, N-methyl-D-aspartate. Adapted from © National Institute of Aging, National Institutes of Health, 2011. All rights reserved. National Institute of Aging, National Institutes of Health [1]. Adapted from © Forest Pharmaceuticals, 2011. All rights reserved. Forest Pharmaceuticals [2].

severe [4,5]. Cholinergic deficits are not found as often in patients with mild AD [6].

By inhibiting acetylcholinesterase, the enzyme that degrades acetylcholine, cholinesterase inhibitors help to increase acetylcholine and, by extension, cognitive function [5]. Rivastigmine and galantamine are approved for use in mild-to-moderate AD and their use generally leads to a 6-month improvement in symptoms [7,8]. However, cognitive function has been shown to decline over time with use of galantamine therapy, and its use over the long term is debatable. One study found that patients who were still responsive to galantamine after 6 months were more likely to benefit from long-term therapy than those who had rapid cognitive decline at that timepoint [9]. No long-term (>26-week), randomized, placebo-controlled studies of rivastigmine have been conducted [8].

Donepezil is the newest cholinesterase inhibitor and the only one approved in the US for the treatment of moderate-to-severe AD [3,5]. It can be given at doses up to 10 mg/day in patients with mild-to-moderate AD and up to 23 mg/day (once-daily sustained-release formulation) in patients with more severe disease [3,5,10].

N-methyl-D-aspartate-receptor antagonists

It has been hypothesized that the persistent activation of NMDA receptors in the brain and excessive calcium influx caused by the excitatory amino acid glutamate contribute to the symptomatology of AD by eventually causing permanent damage [11,12]. Therefore, blocking this channel should moderate this effect of glutamate on neurodegeneration [11,13].

Memantine is effective and well tolerated in patients with moderate-to-severe dementia in AD. It is only approved in the US for this patient population [2]. Clinical studies of its use in mild dementia did not prove efficacy, and a proposal for its approval for mild AD was rejected by the US Food and Drug Administration (FDA) in 2005 [7,14].

The combination of memantine and a cholinesterase inhibitor has been studied in several randomized clinical trials of patients with mild-to-moderate dementia, but its early promising effects could not be sustained [3]. However, memantine added to an already-existing donezepil treatment regimen does have potential in moderate-to-severe dementia [15,16].

Drugs in development

There are many agents currently being tested in human trials, encompassing both symptomatic and disease-modifying treatments. While several promising compounds have failed to meet Phase III clinical trial endpoints in recent years, the following are still in Phase III testing:

- Solanezumab: humanized monoclonal antibody that binds to the central amyloid beta (Aβ) region. Two Phase III, double-blind, placebo-controlled trials, EXPEDITION1 and EXPEDITION2, did not meet cognitive and function primary endpoints but did, however, find a statistically significant slowing of cognitive decline in a pre-specified secondary analysis of pooled data in patients with mild AD [17]. The Phase III EXPEDITION3 trial of solanezumab in patients with mild AD was also initiated in the latter half of 2013 [18].

- Tramiprosate (homotaurine): compound that binds to soluble Aβ to help inhibit amyloid fibril aggregation. Results from the Phase III, placebo-controlled, double-blind Alphase trial did not show significant differences between tramiprosate and placebo, but post-hoc analyses did note a statistical trend toward reduced cognitive decline [19]. Its development potential will probably be as a nutraceutical rather than a drug.

- Intravenous immunoglobulin 10%: a Phase III trial in patients with mild-to-moderate AD was completed in March 2013 [20].

- Gantenerumab: monoclonal antibody currently in a Phase III, multicenter, randomized, double-blind, placebo-controlled, parallel-group study, examining cognition and function in patients with prodromal AD and the safety and pharmacokinetics of subcutaneous injections of gantenerumab or placebo. Select patients will undergo positron emission tomography (PET) scanning to asses brain amyloid. The study is projected to be completed in 2016 [21].

- Nilvadipine: calcium channel blocker approved in Europe for the treatment of hypertension. Phase III trials are currently ongoing in Europe.

- The β-amyloid precursor protein site-cleaving enzyme (BACE) inhibitors: agents that block BACE to prevent Aβ build up. Novel, oral investigational BACE inhibitor, MK-8931, is currently being evaluated in the Phase II/III EPOCH study to assess its safety and efficacy compared with placebo in patients with mild-to-moderate AD [22]. Additionally, a new Phase III study will initiate dosing (the APECS study) to evaluate MK-8931 in patients with amnestic mild cognitive impairment due to Alzheimer's disease, also known as prodromal Alzheimer's disease [23]. There are also Phase I clinical studies with E2609, which have been conducted, with positive interim results showing reductions in Aβ levels in healthy subjects. These results were presented at the Alzheimer's Association International Conference (AAIC) 2012 and the studies are ongoing [24].

In addition, genetics and biomarkers for prediction and detection of early and presymptomatic disease are aggressively being pursued to enable researchers and clinicians to identify and develop treatments for prevention of dementia. The first amyloid positron emission tomography (PET) imaging agent, florbetapir F18 injection, was approved by the FDA in April 2012; other imaging agents are currently in clinical trials [25,26].

References

1 National Institute on Aging. Alzheimer's disease medications fact sheet. Updated November 2012. NIH Publication No. 08–3431. www.nia.nih.gov/sites/default/files/alzheimers_disease_medications_fact_sheet_1.pdf. Accessed November 20, 2014.

2 Namenda [package insert]. St. Louis, MO; Forest Pharmaceuticals, Inc.; 2011.

3 Popp J, Arlt S. Pharmacological treatment of dementia and mild cognitive impairment due to Alzheimer's disease. *Curr Opin Psychiatry*. 2011;24:556-561.

4 Perry EK, Tomlinson BE, Blessed G, Bergmann K, Gibson PH, Perry RH. Correlation of cholinergic abnormalities with senile plaques and mental test scores in senile dementia. *BMJ*. 1978;2:1457-1459.

5 Sabbagh M, Cummings J. Progressive cholinergic decline in Alzheimer's disease: consideration for treatment with donepezil 23 mg in patients with moderate to severe symptomatology. *BMC Neurol*. 2011;11:21.

6 Davis KL, Mohs RC, Marin D, et al. Cholinergic markers in elderly patients with early signs of Alzheimer disease. *JAMA*. 1999;281:1401-1406.

7 Neugroschl J, Sano M. Current treatment and recent clinical research in Alzheimer's disease. *Mt Sinai J Med*. 2010;77:3-16.

8 Birks J, Grimley Evans J, Iakovidou V, Tsolaki M, Holt FE. Rivastigmine for Alzheimer's disease. *Cochrane Database Syst Rev*. 2009;(2):CD001191.

9 Kavanagh S, Howe I, Brashear HR, et al. Long-term response to galantamine in relation to short-term efficacy data: pooled analysis in patients with mild to moderate Alzheimer's disease. *Curr Alzheimer Res*. 2011;8:175-186.

10 Aricept [package insert]. Woodcliff Lake, NJ; Eisai Inc.; 2012.

11 McKeage K. Memantine: a review of its use in moderate to severe Alzheimer's disease. *CNS Drugs*. 2009;23:881-897.

12 Danysz W, Parsons CG, Möbius H-J, Stöffler A, Quack G. Neuroprotective and symptomalogical action of memantine relevant for Alzheimer's disease – a unified glutamatergic hypothesis on the mechanism of action. *Neurotox Res*. 2000;2:85-97.

13 Wenk GL, Parsons CG, Danysz W. Potential role of N-methyl-D-aspartate receptors as executors of neurodegeneration resulting from diverse insults: focus on memantine. *Behav Pharmacol*. 2006;17:411-424.

14 Schneider LS, Dagerman KS, Higgins JPT, McShane R. Lack of evidence for the efficacy of memantine in mild Alzheimer disease. *Arch Neurol*. 2011;68:991-998.

15 Tariot PN, Farlow MR, Grossberg GT, Graham SM, McDonald S, Gergel I; for the Memantine Study Group. Memantine treatment in patients with moderate to severe Alzheimer disease already receiving donepezil: a randomized controlled trial. *JAMA*. 2004;291:317-324.

16 van Dyck CH, Schmitt FA, Olin JT; for the Memantine MEM-MD-02 Study Group. A responder analysis of memantine treatment in patients with Alzheimer disease maintained on donepezil. *Am J Geriatr Psychiatry*. 2006;14:428-437.

17 Lilly Provides Update on Next Steps for Solanezumab [press release]. Indianapolis, IN: Eli Lilly and Company/PR Newswire; December 12, 2012. files.shareholder.com/downloads/LLY/2252462413x0x621344/127181b5-dd7a-4540-b688-c29634ff75b4/LLY_News_2012_12_12_Product.pdf. Accessed November 20, 2014.

18 Progress of Mild Alzheimer's Disease in Participants on Solanezumab Versus Placebo (EXPEDITION 3). http://clinicaltrials.gov/ct2/show/NCT01900665. Last updated October 10, 2014. Accessed November 20, 2014.

19 Aisen PS, Gauthier S, Ferris SH, et al; for the Alphase group. Tramiprosate in mild-to-moderate Alzheimer's disease — a randomized, double-blind, placebo-controlled, multi-centre study (the Alphase Study). *Arch Med Sci*. 2011;7:102-111.

20 A phase 3 study evaluating safety and effectiveness of immune globulin intravenous (IGIV 10%) for the treatment of mild-to-moderate Alzheimer's disease. www.clinicaltrials.gov/ct2/show/NCT00818662. Last updated October 23, 2014. Accessed November 20, 2014.

21 A study of gantenerumab in patients with prodromal Alzheimer's disease. www.clinicaltrials.gov/ct2/show/NCT01224106. Last updated November 17, 2014. Accessed November 20, 2014.

22 Merck initiates Phase II/III study of investigational BACE inhibitor, MK-8931, for treatment of Alzheimer's disease [press release]. December 3, 2012. www.mercknewsroom.com/press-release/research-and-development-news/merck-initiates-phase-iiiii-study-investigational-bace-i. Accessed November 20, 2014.

23 Merck advances development program for investigational Alzheimer's disease therapy, MK-8931 [press release]. December 10, 2013. http://www.mercknewsroom.com/news-release/prescription-medicine-news/merck-advances-development-program-investigational-alzheimer. Accessed November 20, 2014.

24 Eisai presents first clinical data for BACE inhibitor E2609 at Alzheimer's Association International Conference 2012 [press release]. July 19, 2012. www.eisai.com/news/news201247.html. Accessed November 20, 2014.

25 Assess the prognostic usefulness of flutemetamol (18F) Injection for identifying subjects with amnestic mild cognitive impairment who will convert to clinically probable Alzheimer's disease. www.clinicaltrials.gov/ct2/show/NCT01028053?term=Flutemetamol&rank=1&submit_fld_opt. Last updated September 3, 2014. Accessed November 20, 2014.

26 Effectiveness of an electronic training program for orienting and interpreting [18F] flutemetamol positron emission tomography (PET) images. www.clinicaltrials.gov/ct2/show/NCT01672827?term=Flutemetamol&rank=2&submit_fld_opt=. Last updated December 18, 2013. Accessed November 20, 2014.

Chapter 4

Pharmacological treatment of behavioral and psychological symptoms of Alzheimer's disease

Development of behavioral and psychological symptoms of Alzheimer's disease

Behavioral and psychological symptoms of dementia (BPSD) occur in nearly all of patients with dementia during the course of their illness; BPSDs include depression, apathy, psychosis, delusions, hallucinations, agitation, dysphoria, anxiety, disinhibition, irritability, social withdrawal, paranoia, suicidal ideation, mood changes, sexually inappropriate conduct, aggression, wandering, accusatory language, and aberrant motor behavior [1,2]. Some studies indicate that BPSDs may be associated with progression of dementia and occur more often in the moderate-to-severe stages of the illness [1–3]; however, the link between cognitive decline and behavioral symptoms has been difficult to demonstrate.

The type of BPSD may vary depending on the region of the brain affected. Functional brain imaging, single-photon emission computerized tomography (SPECT), and positron emission tomography (PET) imaging have been valuable in defining the relationship between symptom development and underlying pathophysiology, though their use in clinical practice remains limited. For example, in one study utilizing SPECT, investigators found significant hypoperfusion of the left anterior temporal cortex and the right and left dorsolateral frontal cortex in patients with dementia displaying aggression and agitation compared with patients with dementia not displaying BPSD [4]. Psychotic symptoms were associated with

© Springer Healthcare 2015
A. Burke et al., *Pocket Reference to Alzheimer's Disease Management*,
DOI 10.1007/978-1-910315-22-4_4

hyperperfusion in frontal regions combined with hypoperfusion in the posterior temporal region, frontal hypometabolism, severe parietotemporal metabolic deficits, temporal deficits, and left medial occipital and inferior temporal gyrus hypometabolism [4]. Similarly, SPECT studies in patients with AD suffering from apathy have revealed frontal and cingulated hypoperfusion [4].

Neurodegenerative changes present in the brains of individuals with AD result in reduced levels of acetylcholine, serotonin, norepinephrine, somatostatin, and excitatory amino acids. Cholinergic deficiency in the limbic and paralimbic regions is believed to correlate with the development of BPSD [5]. It has also been implicated in sleep disturbances experienced by many individuals with AD. Since a cholinergic deficiency appears to underlie the development of BPSD, cholinesterase inhibitors are likely to ameliorate AD-related behavioral disturbances [5].

Treatment of BPSD should include a thorough assessment of possible underlying causes or exacerbating factors (Table 4.1), because alleviating these factors may minimize the need for pharmacotherapy, hospitalization, and institutionalization (see Chapter 5).

Managing behavioral and psychological symptoms of Alzheimer's disease

BPSDs present a unique challenge with regards to pharmacological treatment for AD as currently there are no treatments approved by

Causes	Cautions
Unrecognized infections	Urinary tract infections in particular
Medication regimen	Check for drugs that may cause or aggravate symptoms
Electrolyte disturbances	Hyponatremia and dehydration may produce confusion/delirium
Constipation	Pain and discomfort due to untreated constipation may lead to distress; check underlying cause, including drugs
Pain	Unrecognized or untreated pain is common in the elderly and is often difficult to identify and assess in a person with dementia
Hearing or vision problems	Make regular assessment of sensory function
Environmental factors	Noise, poor lighting, frustration finding facilities (ie, bathroom, and other environmental factors) can cause distress

Table 4.1 Underlying causes of behavioral and psychological symptoms of dementia.
Reproduced with permission from © Elsevier, 2001. All rights reserved. Cohen-Mansfield [6].

the US Food and Drug Administration (FDA) for management of these behaviors; however, consensus guidelines recommend general treatment plans based on available clinical data to manage specific BPSDs with appropriate pharmacological agents (Table 4.2) [7].

Antipsychotics

As mentioned in Table 4.2, general guidelines recommend the use of antipsychotics when managing a patient with AD and aggression, agitation, and/or psychosis [7]. These authors [8] recommend that before an antipsychotic is prescribed, the benefits and risks of treatment should be assessed and reviewed with the patient, family, and caregivers; specifically, the patient, family members, and caregivers should be made aware of the increased risk of mortality that antipsychotics pose to elderly patients (details regarding atypical antipsychotics for elderly patients with AD are described in Table 4.3). An antipsychotic should only be indicated if aggression, agitation, and/or psychotic symptoms cause severe distress or an immediate risk of harm to the patient or others, or if more conservative treatments have failed. Nonpharmacological measures should be employed concurrently with drug treatments, and pharmacological treatment should be aimed at the modification of clearly identified and documented target behaviors. Regarding dosing, these authors recommend that the patient should be:

- started on the lowest possible dose, and if a dose increase is necessary, it should be titrated slowly to effect; and

Use antidementia agents first
- AChEIs have been shown to have an anti-agitation effect
- Memantine trial showed antiagitation effect

Atypical antipsychotics: first line for psychosis with or without agitation

No first-line recommendation for agitation without psychosis: consider antipsychotic alone or with other agent, or other agent alone
- Mood stabilizers
- Serotonergic compounds
  - Trazodone: negative trials but positive clinical experience
  - Sertraline: anecdotal evidence only
  - Citalopram: preliminary evidence for possible effect
  - Escitalopram: by inference

Table 4.2 Consensus guidelines for the treatment of behavioral and psychological symptoms in Alzheimer's disease, 2004. AChEIs, acetylcholinesterase inhibitors. Adapted from © Physicians Postgraduate Press, 2004. All rights reserved. Alexopoulos et al [7].

Generic	Aripiprazole	Olanzapine
Elderly dosing	**Initial:** 5–7.5 mg/day	**Initial:** 2.5–5 mg/night
	Titration: May be increased to a maximum of 30 mg/day	**Titration:** Increase by 5–10 mg/day at 1-week intervals; maximum dose: 20 mg/day
Side effects	Sedation (+), weight gain (±), orthostatic hypotension (±), EPS (+), Parkinsonism (>2%), akathisia (>10%), dystonic prescriptions (<2%). Other side effects: headache, anxiety, insomnia, nausea	Sedation (++), weight gain (++++), orthostatic hypotension (++), EPS (++), Parkinsonism (>2%), akathisia (>10%), dystonic prescriptions (<2%). Other side effects: hyperglycemia, diabetic ketoacidosis

Table 4.3 Atypical antipsychotics (continues overleaf).

- regularly reviewed (after initial follow-up and then every 3 months) for clinical response and adverse effects.

If the patient needs to stop or decrease their treatment, these authors recommend that withdrawal of antipsychotics should be done gradually by reducing the dose by 50% every 2 weeks, and then stopping after 2 weeks on the minimum dose, with monitoring for recurrence of target symptoms or emergence of new ones. The longer a medication has been prescribed, the slower the withdrawal should occur; thus, there will be less possibility of emerging symptoms related to drug withdrawal.

Depression

Diagnosis of depression in AD can be difficult. Some of the problems with the diagnosis arise because patients with AD may not be able to describe their feelings, their experience of the way they feel, or give an accurate history of their mood changes or state. Diagnoses are frequently made based on caregiver observations. Seniors may not meet full *Diagnostic and Statistical Manual of Mental Disorders, 5th Edition* criteria for a major depressive disorder since they often tend to present with more somatic complaints [8]. Many symptoms of depression (Table 4.4), such as insomnia, fatigue, and difficulty with concentration, may be attributed by clinicians as being related to comorbid medical conditions and dementia [8].

Quetiapine	Risperidone	Ziprasidone
Initial: 12.5–25 mg twice daily **Titration:** Increase by 25–50 mg twice daily every 1–2 days; maximum dose: 400 mg/day	**Initial:** 0.5 mg/twice daily **Titration:** Increase by 0.5–1 mg/day every 1–3 days; target dose: 4–5 mg/day	**Initial:** 20 mg twice daily with food **Titration:** Increase by 20 mg twice daily every 2–3 days to 60–80 mg twice daily with food
Sedation (++), weight gain (++), orthostatic hypotension (++), EPS (±), Parkinsonism (>2%), akathisia (>2%), dystonic prescriptions (<2%). No known additional side effects	Sedation (+), weight gain (++), orthostatic hypotension (++), EPS (++), Parkinsonism (>10%), akathisia (>10%). EPS and hyperprolactinemia generally occur at doses >6 mg	Sedation (++), weight gain (±), orthostatic hypotension (+), EPS (++), Parkinsonism (>2%), akathisia (>2%), dystonic prescriptions (>2%). Contraindicated in recent acute myocardial infarction, persistent QTc >500 ms

Table 4.3 Atypical antipsychotics (continued). EPS, electrophysiological symptoms.

Treatment of depressive symptoms in AD should include a thorough evaluation of potential environmental and psychosocial triggers, which can exacerbate the condition (see Chapter 5 for more information). If pharmacotherapy is warranted, these authors recommend selective serotonin reuptake inhibitors (SSRI) as first-line therapy [8]; also, monoamine oxidase inhibitors and tricyclic antidepressants should be avoided due to their adverse effects on cognition, as well as a greater potential for other serious adverse effects (Table 4.5).

Unexplained or aggravated aches and pains
Hopelessness and/or helplessness
Anxiety and worries
Memory problems
Apathy
Increased use of substances
Fixation on death
Loss of feeling of pleasure
Slowed movement
Irritability
Lack of interest in personal care (skipping meals, forgetting medications, neglecting personal hygiene)
Guilt/worthlessness

Table 4.4 Symptoms of late-life depression. Adapted from © American Psychiatric Association, 2013. All rights reserved. American Psychiatric Association [8].

Drug class	SSRI	SNRI
Specific drug(s), dosage	Citalopram 20–40 mg/day	Venlafaxine 75 mg/twice daily
	Escitalopram 10–20 mg/day	Venlafaxine XR 75–225 mg/day
	Sertraline 50–200 mg/morning	Duloxetine 30–120 mg/day
	Fluoxetine 20–40 mg/morning	
Side effects and warnings	Nausea and diarrhea might occur	Fewer drug interactions
	Sedation or activation may occur	Can cause or aggravate hypertension
	SIADH at high doses and sexual side effects	Risk for withdrawal syndrome
	Interact with CYP-450 isoenzymes by inhibition	
	Can increase the anticoagulant effect of warfarin	
	Do not discontinue abruptly; taper the dose	

Table 4.5 Pharmacotherapy for depression for patients with Alzheimer's disease (continues over). CYP-450, cytochrome P450; SIADH, syndrome of inappropriate antidiuretic hormone; SNRI, serotonin–norepinephrine reuptake inhibitor; SR, sustained-release; SSRI, selective serotonin reuptake inhibitors; XR, extended-release.

Apathy

Apathy is a disorder of motivation that is characterized by reduced goal-directed activity in domains of behavior, cognition, and emotion. Apathy is a frequent symptom of many neuropsychiatric conditions. Clinical features of apathy generally present through:

- behaviors: eg, lack of effort, productivity, structure, initiative;
- cognitions: eg, diminished interest, curiosity, concern, insight, planning, goal setting; and
- emotions: eg, decreased emotional responsiveness to both positive and negative events, flat affect.

Pharmacological approaches to apathy include cholinesterase inhibitors, dopaminergic antidepressants (eg, sertraline, bupropion), stimulants (eg, methylphenidate, dextroamphetamine), and dopamine agonists (if electrophysiological symptoms are present). However, unlike depression, apathy is less likely to respond to pharmacological interventions and the effective approaches are environmental and behavioral modifications, which may affect the treatment of a patient with AD, suffering from depression or apathy (see Chapter 5).

Norepinephrine, 5HT$_2$ and 5HT$_3$ antagonist	Norepinephrine-dopamine reuptake inhibitor
Mirtazapine 15–45 mg/nightly	Bupropion 100 mg/three times daily
	Bupropion SR 150 mg/twice daily
Can cause serotonin syndrome when given with other SSRI	Less sexual dysfunction
Frequently used due to potential benefits on sleep, appetite as well as anti-emetic properties	

Table 4.5 Pharmacotherapy for depression for patients with Alzheimer's disease (continued). CYP-450, cytochrome P450; SIADH, syndrome of inappropriate antidiuretic hormone; SNRI, serotonin–norepinephrine reuptake inhibitor; SR, sustained-release; SSRI, selective serotonin reuptake inhibitors; XR, extended-release.

Insomnia

Sleep disorders frequently affect those suffering from AD. Degeneration of nerve cells in the circadian pacemaker in the brain is thought to be responsible for many of these sleep disturbances; however, other factors may also play an important role. For example, medications used to treat dementia can adversely affect sleep.

Treatment of insomnia in dementia should begin with a thorough evaluation of environmental factors and physical conditions, which may be adversely affecting the patient's ability to obtain a restful night's sleep. Pharmacological approaches to insomnia for patients with AD are listed in Table 4.6, and are based on the authors' clinical experience.

Antidepressants
May minimize anxiety, irritability, and agitated behaviors
Can improve daytime energy level, interest, and apathy
Side effects: gastrointestinal upset, jitteriness, activation, sedation, headaches, vivid dreams, 'hangover'-like feelings
Some antidepressants are more sedating (eg, trazodone, mirtazapine) than others
Avoid: amitriptyline, paroxetine
Hypnotics
May be of some benefit in individuals having difficulty falling and staying asleep
Can increase confusion in people with dementia
Can cause 'hypnotic' states (can result in rebound insomnia)
Antipsychotics
Typically not used as 'sleep aids' due to potentially serious side effects
Used to treat agitation related to sundowning behaviors
Have anti-anxiety, anti-agitation, and sedative properties
Side effects: sedation, worsening mobility (electrophysiological symptoms), falls, orthostatic hypotension, leg swelling, increased mortality

Table 4.6 Possible pharmacological treatment approaches of insomnia for patients with Alzheimer's disease. Based on the authors' clinical experience, they recommend to avoid antidepressants: amitriptyline, paroxetine, and benzodiazepines, which may worsen confusion, are addictive (can results in rebound insomnia), may increase the risk of falls and fractures due to sedative properties, and can result in paradoxical responses/worsen sundowning in patients with dementia.

References

1 Cummings JL, Fairbanks L, Masterman DL. Strategies for analysing behavioural data in clinical trials involving patients with Alzheimer's disease. *Int J Neuropsychopharmacol.* 1999;2:59-66.

2 Robert P. Understanding and managing behavioural symptoms in Alzheimer's disease and related dementias: focus on rivastigmine. *Curr Med Res Opin.* 2002;18:156-171.

3 Jost BC, Grossberg GT. The evolution of psychiatric symptoms in Alzheimer's disease: a natural history study. *J Am Geriatr Soc.* 1996;44:1078-1081.

4 Hirono N, Mega MS, Dinov ID, Mishkin F, Cummings JL. Left frontotemporal hypoperfusion is associated with aggression in patients with dementia. *Arch Neurol.* 2000;57:861-866.

5 Burke A. Pathophysiology of behavioral and psychological disturbances in dementia. In: McNamara P, ed. *Dementia. Vol. 2: Science and Biology*. Santa Barbara, CA; ABC-CLIO, LLC; 2011:135-158.

6 Cohen-Mansfield J. Nonpharmacologic interventions for inappropriate behaviors in dementia: a review, summary, and critique. *Am J Geriatr Psychiatry.* 2001;9:361-381

7 Alexopoulos GS, Streim J, Carpenter D, Docherty JP; Expert Consensus Panel for Using Antipsychotic Drugs in Older Patients. Using antipsychotic agents in older patients. *J Clin Psychiatry.* 2004;65(suppl 2):5-104.

8 *American Psychiatric Association. Diagnostic and Statistical Manual of Mental Disorders (DSM-5).* Arlington, VA: American Psychiatric Association; 2013.

PART TWO

Caring for patients with Alzheimer's disease and related dementias

Chapter 5

Managing patients with Alzheimer's disease and related dementias

Management is one of the most common reasons why families seek medical attention for people with dementia. Families often know that the person has Alzheimer's disease (AD), but need information on how to manage dementia and its symptoms after the diagnosis is established. Issues vary by stage, but may include repetitive behaviors, hiding objects, night awakening, irritability, refusal to bathe, wandering, late-day confusion, self-centeredness, nonrecognition of home or caregiver, and many others, all of which may heighten family conflict. Many of these symptoms are not well managed by medications, and their use may actually worsen the situation. Physicians need to understand the principles of dementia management in order to counsel families and suggest meaningful resources for day-to-day care.

AD is the most common form of dementia; therefore, most studies of nonpharmacologic interventions have been conducted in people with AD. The following sections refer mostly to Alzheimer's-type dementia (AD dementia), and examine disease presentation and evolution of AD dementia in terms of symptoms, solutions, and family needs by stages; these sections may not work as well with non-Alzheimer dementias, such as frontotemporal dementias or Lewy body disease.

Care issues in mild cognitive impairment and mild dementia

There are many important legal, financial, and health care decisions to be made early in the disease, especially when the person has the ability

© Springer Healthcare 2015
A. Burke et al., *Pocket Reference to Alzheimer's Disease Management*,
DOI 10.1007/978-1-910315-22-4_5

to provide input and preferences. Safety issues commonly arise in situations (eg, driving and living alone) and they must be addressed as soon as possible as these problems will increase as the dementia advances with time. Daily living can be optimized through good health maintenance, a focus on function, and thoughtful planning and input from the clinician. Caregivers often report that persons within early stages of dementia have these behavioral problems listed below; the person may:

- lose/misplace things;
- repeat stories;
- forget recent conversations;
- mislabel names/words;
- dislike social situations;
- refuse to give up cherished activities (eg, driving);
- more likely argue and get frustrated;
- complain that caregiver is 'controlling;'
- accuse spouse of infidelity or acquaintances of stealing money;
- forget to pay bills;
- mismanage medications;
- not show interest in previously enjoyed activities;
- diminish sense of risk, express poor judgment;
- confabulate stories in general and from things seen on television;
- show signs of paranoia, hoarding, social isolation when living alone; and
- be at risk for exploitation.

Starting the discussion

The prospect of discussing the probable bleak future for patients with AD may be frightening for the patient, their family members, and caregivers. However, clinicians should encourage these discussions as fears may be allayed by early conversations between the patient and their caregivers. Also, by having a dialogue early in the illness about concerns and long-term care needs, the patient will be able to participate, allowing them to express both their preferences and their fears before the disease develops further. Some of the suggested topics that may be discussed with the patient include the following:

- If the person with dementia became confused and unable to manage decisions about their health and/or finances, is there a person they have designated to help them to make those decisions? If that person was not available, do they know who they would want as a backup?
- If the person with dementia has no one in mind to manage decisions, suggest that it is a good idea to find someone. Once a person (or people) has/have been identified, suggest that the person with dementia obtains the permission of that person (or people) before including them in a written plan.
- If the care needs of the person with dementia exceed what can be provided at home, ask them to consider where they would want to be? Do they know who would they want to help them? It is also important to ask the person with dementia if they needed to go into a facility, is there a place they would prefer to go to?

Legal issues

When dementia is first diagnosed, decisions must be made and documents finalized well in advance to prepare for the person's incapacity. Laws concerning these documents vary, but the documents are usually referred to as 'durable' or 'lasting' power of attorney, meaning the document contains a clause stating that the power of attorney is still valid if the person becomes incapacitated. In the US, for example, there are two or three types of durable powers of attorney: 1) medical or health care; 2) financial; and 3) mental health. The latter allows a family to pursue acute psychiatric hospitalization without court involvement or standby powers, which require a clinician's statement of incapacity to enact. An attorney who specializes in elder law, family law, or probate law should be consulted to help draw up these documents and a copy of the medical/health care durable power of attorney should be filed in medical and hospital records.

The family can have an attorney file for guardianship and/or conservatorship if the patient refuses to cooperate or is unsafe, if there is serious family conflict, or if there is a history of exploitation, self-neglect, or abuse. These are court-ordered documents that appoint a person for parent-like (guardianship) or financial (conservator) power over the patient and/or the estate. The clinician will be asked to provide documentation about the patient's capacity [1].

Families may try to avoid drawing up these documents; however, failing to prepare for the day when a loved one cannot make decisions can affect the patient, spouse, and other family members and caregivers, and could result in the following:

- bills may not be paid;
- pension/government support problems will be difficult to resolve;
- property and income tax may not be paid appropriately or in a timely manner;
- due to federal privacy laws, the family may be refused access to medical information, including when the person with dementia is hospitalized, has a sudden change in behavior, or has problems with medications [2];
- the person with dementia may be at increased risk for exploitation because he/she will still be able to sign contracts;
- end-of-life decisions and care preferences will not be known and/or implemented;
- banks, financial institutions, insurance companies, and other similar businesses may refuse to work with the family without the presence or permission of the family;
- other relatives or total strangers may come in and obtain a power of attorney when the person with dementia does not understand what they are signing; and
- family members may become deeply divided (eg, arguing about what they feel is the best care for the person with dementia).

Married couples often believe that marriage entitles them to information about their spouse or gives them the right to make legal, financial, or health care decisions on their loved one's behalf. However, federal laws on privacy affect what kind of health care, insurance, banking, and investment information can be disclosed; spouses are not automatically entitled to such information or designated as substitute decision-makers. Therefore, if preparations are not made well in advance, the result could be prolonged and costly legal actions, with the patient's wishes often remaining unknown or disregarded [3].

Since caregivers are also at risk for becoming ill and may predecease the patient, both the person with dementia and their caregivers should complete advanced directives.

Driving

Driving ability requires immediate attention once a person has recognizable memory loss, with continued scrutiny as long as the person drives. Driving is a complex activity that begins to decline very early in the course of AD in the following ways:

- The ability to see or perceive changes:
  - depth perception (the ability to judge distances): may result in following too closely or too far, having dents in the car from sideswiping objects, mishaps with distance, and rear-end collisions;
  - see moving objects: the person does not see things moving across the horizon or towards them, which may result in not seeing oncoming traffic, cars at intersections, trains, or children darting across a road; and
  - sense of verticality
- The ability to recognize familiar places: may occur spontaneously while driving or when a mildly unexpected event occurs; examples include asking directions or getting lost in familiar places, making wrong turns, having difficulty following directions from a passenger, or getting lost after a relatively minor event.
- The ability to remember: person does not remember where they came from, why they are in the car, or how to get to where they were planning to go. Later in the illness, they often forget how to operate the car as evidenced, for example, by hitting the gas instead of the brake pedal.
- The ability to change body posture quickly: staying in an upright position (eg, after turning a corner), controlling the vehicle..
- The ability to make sound judgments quickly:
  - respond negatively to being cut-off by another driver: the person may not respond well and may retaliate, become confused, or cry;
  - respond negatively to heavy traffic;
  - respond negatively to being rear-ended;

- – behave inappropriately at a train crossing;
- – make a 'left on red;'
- – react to an accident: the person may not realise they have had an accident or may leave the scene after having an accident;
- – try to pass slower vehicles by crossing lanes into oncoming traffic or a no-passing lane;
- – react to a child running in front of the car: the person may not be able to stop quickly;
- – respond inappropriately to emergency vehicles;
- – be unable to manage of 'near miss' situations;
- – be unable to cope with the speed limits on highways and on city streets for which speed limits may be higher;
- – drive erratically and/or stop suddenly at inappropriate places;
- – become 'befuddled' in situations that are 'beyond the norm;' and
- – respond inappropriately if they become lost

Emotional responses to driving

It is important to recognize that there are few privileges more cherished than driving. It is a reflection on a person's ability to be independent and autonomous. When the ability to drive is taken away, many people become angry, depressed, and prone to social isolation. Driving is an extremely emotional issue and one where families often strongly disagree on management.

Based on the authors' clinical experience, incidence of accidents both involving and caused by people with dementia rises sharply, even very early in the disease [4]. While the person with dementia may deny the risk, the reality is that they will experience difficulty while driving. Mandatory reporting laws vary and may serve as a disincentive to seek medical care.

The physician should begin this important discussion as soon as the disease is diagnosed, or even earlier if problems are noticed. The first discussions should center on the person's illness; for example, the discussion could begin with: *"You know you have an illness called dementia. This means the day is coming when you will have to give up driving. You may not know when you are unsafe, so someone will have to tell you. Who will you trust to tell you when you need to stop?"* This dialogue informs the person that giving up driving is inevitable and brings them into the decision-making process.

Once the 'decider' is determined and they are told of their responsibilities, ask them to observe the person's driving on a regular basis: monthly if the person is mildly impaired or weekly if the impairment is obvious. As many members of the family as possible should witness this discussion in order to develop agreement among them.

The physician should also suggest that the patient, family members, and caregivers obtain additional safety precautions if the person is still driving:

- Medical identification tag or band: a person with early dementia who continues to drive should have a medical identification tag or band in case they are stopped by police. This helps to assure the person that they will not be arrested for driving under the influence if they are mildly confused.
- GPS devices: there are portable GPS devices that allow a caregiver to instantly locate a patient that has become lost while driving and alert the police. Please see page 92 for useful resources to obtain a GPS device.
- 'No driving' prescription: the physician may want to 'prescribe' a patient not to drive, which the family can then use as a prompt to discuss the patient's options. In difficult cases, the physician may write the 'prescription' and have the designated person and physician sign it, keeping a copy for the medical record.
    - The physician should also mention their obligation under state law to report unsafe drivers;
    - Remind the patient that 'no driving' includes not only driving automobiles, but also trucks, snowmobiles, tractors, and boats and other water vehicles; and
    - The physician should reward any move toward driving cessation with positive feedback;

A follow-up discussion about driving should be undertaken at each visit. This helps the person understand that they will have to completely stop driving soon. However, if the person does not voluntarily relinquish driving, there are several options:

- Discuss it with the person: the person may be angry, which is a normal response to grief and loss, but make sure to listen and support the patient when they talk to you. Apologize that they are feeling upset, but remain steadfast that they cannot drive.

- Explain the danger of lawsuits and legal action: inform the person that he/she (and the provider) could be held liable if they have an accident and that their diagnosis may be disclosed during the course of the lawsuit. Depending on the country's laws, the disclosure of their disease could result in being dropped by insurance or having financial losses.
- Mention the role of government agencies: For example, in the US, a law enforcement office, physician or even an anonymous concerned citizen can report an unsafe driver to the Department of Transportation, which can lead to a suspension of driving privileges.

For those people who insist they are still safe to drive, refuse to give up driving, or if the physician or family is unsure about safety, a driver assessment program or occupational therapy assessment for driving can be helpful. Primary care providers are responsible for counseling the person and family about driving, and failure to do so could lead to litigation; thus, the physician must make sure each discussion is well documented and includes family concerns, the physician's impressions, recommendations, and the patient's response.

Living alone

If a person with dementia is living on their own, rather than with a family member or friend, there are likely to be concerns about their ability to cope, especially as their dementia progresses [5]. The loss of executive function (ie, the inability to plan, initiate, and coordinate the sequence of steps to reach a goal) can lead to excess disability. This is especially true for a person with AD who lives by themselves and does not have someone to assist them when they are trying to execute voluntary activities. Instead, many solitary patients with AD are likely to focus on thinking about the task but are unable to actually complete it. Many early and moderately-staged patients in this circumstance are aware of their loss of executive function, and can be frightened by it as well.

Similarly, attempting to conceptualize time and scheduled events are other obstacles solitary people with AD will face. Since time is abstract and managing appointments is complex, the person may call repeatedly to confirm a meeting time or show up hours early for an appointment. Difficulty with timekeeping issues can cause considerable fear and discomfort for the person, resulting in them giving up previously favored activities, becoming more socially isolated, increasing hoarding behaviors, and developing paranoid ideas.

Families are often concerned about moving the person from their home, believing that 'home is always the best place;' however, this is not always the case in AD-type dementia. Patients can be better off in places where there is structure, support, and other people. In fact, persons who live alone can often have significant functional and mood improvement if they are moved to a good facility. Patients with AD are prone to lose their sense of risk, which can result in a variety of safety issues, especially for those living alone. Some of these risks include home security problems, medication administration issues, weight loss, food poisoning, falls, self-neglect, exploitation, and fires.

The assessment listed in Table 5.1 can help families and clinicians determine whether a person is safe living alone and to recognize when a move should occur.

Care issues in moderate dementia

The moderate stage of dementia may be the most challenging for the caregiver and the most distressing for the patient with AD. Behaviors commonly reported in this stage by caregivers include:

- no sense of time;
- depression;
- wanting to 'go home,' regardless of location;
- becomes agitated with changes of environment;
- nonrecognition of familiar people and caregiver; may think caregiver is an imposter;
- 'clings' to caregiver, refusing to leave their side;
- refuses personal care, becoming aggressive when forced; refuses to bathe;
- tries to wander/escape;
- may wander aimlessly and/or rummage through drawers;

Grade			
A = Emergent	**A/B = Emergent/Semi-emergent**	**B = Semi-emergent**	**C = Non-emergent**
Only **one condition** needs to be present. Immediate help or placement is required	These are situations where, depending on the underlying cause of the problem, could be either A or B	**>2 conditions** indicate that there are safety concerns that must be addressed	**>3 conditions** are present. Additional help will be beneficial Re-evaluate monthly
Observed or reported conditions			
— Weight loss of >6 pounds or 10% body weight in 6 months, evidence of protruding bones	— Malfunctioning plumbing	— Not able to manage bowel/bladder care	— Phone calls from community members; advising help is needed
— Presence of paranoia, hallucinations, delusions, aggression, or thoughts of suicide	— Thermostats not set appropriately for weather conditions	— Repeated calls to family/others asking what to do next	— Vegetative or socially isolated behavior (sitting all day with TV on or off)
— Threatens violence with/without weapons	— Chronic anxiety, panic attacks, worry, or depression is present	— Dirty/infested household	
— Evidence of caregiver injury/domestic violence	— Unsafe driving or refuses to stop driving	— Garbage accumulation	— Missing belongings, hiding things
— Repeated emergency room visits, hospitalizations	— Neighbors calling police	— Food stored inappropriately	— Poor grooming, wearing the same clothing all the time; soiled appearance
— Evidence of substance abuse		— Taken advantage of by family, friends, neighbors	
— Frequent calls to police or emergency services		— Refuses personal care for prolonged period of time	
— Wandering outside the home			
— No food/rancid food in the home			
— Lack of safety with stove, power tools, yard			
— Unable to take medications correctly			
— Live stock/other animals receive inadequate care			
— Eviction notice served			
Total A:	**Total A/B's:**	**Total B:**	**Total C:**

Table 5.1 Live alone assessment. The following conditions may indicate when a person with dementia is no longer safe to live alone or will require more services, assistance, or placement. Adapted from © University of Iowa College of Nursing, 2004. All rights reserved. Hall et al [6].

- may speak or behave inappropriately, including sexual behavior;
- forgets to eat/drink, develops purposeful anorexia;
- may confuse silverware;
- more confused during evening hours (sundowning);
- may sleep >20 hours/day;
- gets days and nights confused as sleep becomes more sporadic;
- easily agitated when unable to express self or complete a familiar task;
- has a 90 minute activity tolerance;
- loses sense of risk/danger; and
- illusions of people in home due to altered perception of television, pictures, mirrors.

Behavioral disturbances are much more frequent as the person experiences increased confusion, frustration in daily living tasks, communication failures, dysregulation of emotion, and mounting disability from this progressive disorder (see Table 2.6). Moderate AD dementia changes visual perception and sense of time and produces sensitivities to the environment and fatigue. This necessitates a different approach to care, including structure control of stimuli and focus on activities.

An evidence-based theoretical care model, Progressively Lowered Stress Threshold, is effective in planning and evaluating care in mild and moderate stages of AD dementia [2]. When used by family members or professional caregivers, it assesses and helps families and caregivers cope with significant declines in noncognitive behavioral symptoms, including night-wakening, sundown confusion, agitation, restlessness, aggression, and psychosis. The model outlines the basic knowledge caregivers need to provide for planning days throughout the illness, and is described in the following sections.

Excess disability

The person with dementia begins to have subtle changes in their ability to plan, tolerate stress, and interpret things in the environment as soon as the first symptoms appear. These changes strain the person's ability to function in a normal manner. Increased strain or frustration will eventually trigger dysfunctional or problem behaviors, referred to as 'excess disability.' This implies that triggers are likely to exacerbate the symptoms of the disease without making the actual disease worse. When caregivers are educated on how to prevent these triggers in daily care, it will result in minimizing or eliminating these unwanted secondary behavioral symptoms.

Most noncognitive behavioral symptoms are stress-induced. The person's ability to tolerate stress diminishes with disease progression. The seven stress triggers which make excess disability symptoms worse are listed below; the following section contains techniques for avoiding these triggers:

- Fatigue: people with dementia tire very easily because they intensely concentrate on common tasks, activities, and time.
- Change: people with dementia have problems planning, as they think about activities more often than actually carrying out the activity.
- Overwhelming/misleading stimuli:
  - Overwhelming stimuli: people with dementia may often get confused in uncomfortable, crowded, or loud settings and may act out with angry outbursts or rude statements; and
  - Misleading stimuli: people with dementia may get confused by television, mirrors, and pictures by believing the people they see in these stimuli are real and/or in their home.
- Loss of meaningful activities: activities are the single most important aspect of dementia care. The primary role of activities is to allow the person with dementia to focus on remaining abilities and strengths, thereby enhancing the person's self-esteem.
- Other people refusing to discuss the illness with the person with dementia: people with dementia are generally aware that something is wrong. Having discussions with the patient results in less frustration, fear, paranoia, and suspiciousness and lets them grieve for their loss.
- Creating too much demand: overexercising and testing patients with dementia can result in frustration and unwanted behaviors.
- Secondary illness: medication reactions, infections, pain, or other physical distress that causes delirium.

Maintaining health and function
Exercise
While the cause of AD is unknown, research suggests that maintaining good overall health may forestall symptom onset and perhaps slow

disease progression. Healthy activities, exercise, optimal management of other diseases, and prevention of infections may help the person with dementia stay active and functioning longer. Therefore, it is very important that the person with dementia and their family pay special attention to routine health care and exercise.

Exercise is a key component in any program for healthy living. Research shows that patients with AD who exercise for at least 20 minutes a minimum of three times a week have better intellectual functioning, improved balance, fewer falls, and may actually increase the amount of brain mass [7]. Research also shows that age does not limit the ability to benefit from exercise. People can benefit from regular exercise well into their 80s and show improvement in heart function [8].

It is recommended that people – even those with dementia – partici-pate in aerobic exercise at a minimum of 3 days a week. Aerobic exercises include walking, gardening, mowing the yard, vacuuming, dusting, dancing, swimming, and other activities that get the body moving [8].

Diet

There are no special dietary regimens or supplements that have been found to help AD or other forms of dementia. In general, a well-balanced diet is recommended, including protein, fruits, vegetables, whole-grains, and some starches (such as potato or pasta); however, this can be difficult to maintain if the person is no longer able to plan meals, shop, or prepare a balanced diet. The patient may benefit from eating at congregate meal sites, as they will be able to eat with others and do not need to prepare their food. Patients with AD may also benefit from:

- multivitamins to maintain nutritional status; however, be aware of how multivitamins affect medications;
- snacking on favorite or high-calorie foods to increase appetite and variety of food intake; and
- five or six small meals each day instead of larger meals.

Fluid intake

Many people with dementia do not drink enough fluids, which can lead to urinary infections, bladder irritation, premature loss of urinary control, and constipation. Clinicians should remind caregivers to frequently offer fluids in any form, with the exception of caffeinated beverages and sports drinks.

Rest

People with dementia often need more rest than before the onset of their illness [2]. The increased need to rest can be accommodated by allowing the person to:

- 'sleep in' until they awaken naturally, which will avoid late-day confusion and nighttime waking;
- nap or have quiet periods of up to 90 minutes in the morning and/or afternoon. If the person naps, have him/her sleep in an easy chair, on the sofa, or on top of a made bed; upon awakening, they are more likely to realize the rest period was a nap rather than the waking from a night's sleep;
- use their 'best time of day' for higher stimulus activities, including social events and appointments;
- plan activities for shorter durations (ie, ≤90 minutes in moderate dementia) and spread activities out over a longer period of time; and
- plan rest times during holiday gatherings, special occasions, or when the person is away from home.

Giving a patient with dementia extra rest can help them maintain energy and orientation for later in the day. As the disease progresses, additional brief naps or rests may be added as the person tires [2]; however, if the person is sleeping too much, the clinician should assess the person for depression, underlying medical illnesses, boredom, and/or an inability to start activities.

Pleasurable activities

People who remain active and engaged in pleasurable activities tend to be functional longer, have less depression, and fewer problems with behavioral symptoms. In fact, activities are thought to be the most important aspect of care for people with dementia. A variety of these activities is important; however, they must be activities enjoyed by the person with dementia. This may include varied and, at times, mildly challenging activities, such as those listed below:

- day-to-day activities (eg, bathing, dressing, eating, cleaning up, and routine chores);
- exercise;
- socialization with accepting and meaningful people to provide a sense of self;
- spiritual and/or religious activities;
- adult day care programs (provides structured day, social contact, and meaningful activities);
- puzzles and other intellectually stimulating activities;
- musical and artistic activities, including attending theatre and concerts;
- social interaction consistent with the person's pre-illness lifestyle;
- enjoyable hobbies or skills, relative to the patient's interests;
- safe home-based activities (eg, attending to pets, gardening, household chores, painting, simple cooking);
- intergenerational activities;
- shopping excursions; and
- travel (with adaptations)

Activities that are enjoyable, safe, and hold the patient's attention should be encouraged. Caregivers should be encouraged to obtain books about activities for people with dementia; please see page 93 for a resource list of recommended books and activities.

It is important for clinicians to remind caregivers and family members that as dementia progresses, the time engaged in activities will be shorter. If the person's favorite activities become too difficult, caregivers should either simplify the activity or find new ones to take their place. Also, while doing activities, the patient should be given a choice whenever possible, and not simply 'yes/no' options. When the patient has an option, they may be more reasonable and willing to do the activities, instead of simply saying 'no' out of fear of change. Occupational therapists, recreational therapists, and adult day programs can be sources of help in modifying or finding new activities.

As patients with dementia have difficulty accepting change, it is equally important to have enjoyable routines in consistent patterns that they follow during the day. While the exact timing of the routine is not important, the sequence of activities is very important to in order keep the patient in a familiar environment [2].

Likewise, caregivers should take caution when redecorating the house, decorating for holidays, moving, or even rearranging the furniture as it may cause increased confusion, resulting in noncognitive behaviors. Holidays should be kept simple and quiet, as this will more likely help to avoid behavioral problems from the person with AD. To avoid confusion or disputes, family members should announce plans or activities at the last possible moment, as planning in advance may lead to agitation and refusal due to the patient's fear of change. Travel should be planned to incorporate routines, best times of day, and rest periods. Should the patient get lost, identification bracelets or GPS location devices (see page 92) should be used; however, people with moderate AD should never travel unaccompanied.

Television and films

Watching television or films are not considered to be meaningful activities, and during the moderate stages of the disease may actually trigger illusions and delusions, as can mirrors and pictures. At times, though, the use of television for programs that are special or appealing to early-stage patients may be acceptable in moderation (eg, favorite sporting events, comedies, game shows, or musicals). Similarly, playing old favorite movies and TV shows can be a comforting source of entertainment; however, it is important to note that the person with moderate-stage dementia might think that what they saw on television or film happened in their home. A good rule for family members to determine which programs should be avoided is to consider which television or movie characters they would not want to have in their home, as the patient might imagine such a scenario.

It is recommended that caregivers of patients with moderate-stage dementia should remove family pictures, cover mirrors, and cover the windows at night to reduce confusion. Also, caregivers may have better outcomes with their loved ones if less-graphic television shows and films are shown; more sexually suggestive or violent media may result in confusion, paranoia, and angry outbursts. For instance, while sporting events can be a good source of entertainment, sports that focus on human contact, such as wrestling, boxing, or extreme sports, may provoke a reaction that becomes dangerous for the patient and caregiver alike. Ultimately, television should never be the primary source of activities [2].

Communication

Patients with early-stage AD are often fully aware that something is wrong, and it is necessary for clinicians and caregivers to let these patients discuss their frustrations and confusions. Caregivers should be encouraged to allow the person time to talk about their losses, as grieving is normal and even desirable for the person. Discussing the disease process will help the person understand that they are not 'going crazy.' Additionally, fears of abandonment can be allayed by reassuring the person that their caregivers intend to see and care for them for the foreseeable future.

However, the patient may become angry or deny their memory loss. If this happens, caregivers should be instructed to drop the subject for a time. The goal is not to get the person to admit the memory loss, but to help them understand why it is happening. Denial and anger are a normal part of the grieving process. At the same time, it is not helpful for the caregiver to assume that ignorance of the disease process will lead to a happier patient.

Additionally, the family should be consulted to contact the clinician if the person develops a low mood that lasts for more than 3 weeks and affects sleep or appetite. Depression is a common occurrence in dementia and is a common cause of excess disability. Failing to treat depression worsens dementia symptoms and decreases the person's quality of life [9].

Demands on the patient

Many caregivers feel they need to exercise the brain of a person with AD, testing the person every day and pushing them to achieve. These testing techniques, while done with good intentions, are likely to create unwanted behaviors. Instead, clinicians should reminds caregivers that AD is a disability, and they should assume that the person with dementia is working as hard as they can to live through the disability. Also, many people with early-stage dementia are aware of their disability; thus, constant testing questions (eg, "Do you remember me?", "What is their name?", "Remember what we did yesterday?") may only result in making the person feel frustrated and as if they are failing. It is also helpful to remind caregivers that persons with dementia routinely have good and bad days, which could vary by the hour; the caregivers must learn to accept the person's changing abilities to help manage their AD.

Instead of confronting a patient about their memory loss, caregivers may find that distracting the patient can be an effective technique when working through the person's memory loss. Humor can be an effective tool, and caregivers should be encouraged to use humor and other distractions repeatedly if they are helpful. Although, if the person has forgotten how to do an activity for example, caregivers should assist them by talking them through the activity one step at a time.

Preventive care and immunizations
Delirium and infections
Delirium, or a sudden change in memory, thinking, or behavior, is one of the first signs of illness in someone with dementia. Common causes of delirium include infections, medication reactions, sudden illnesses, pain, and poorly controlled chronic/ongoing illnesses. Unrelenting episodes of sudden confusion should be evaluated medically as soon as possible [2].

Urinary tract infections are one of the more common types of infections that cause delirium. Often people with dementia do not drink enough fluids, may not change incontinence products, or may not be able to cleanse themselves adequately following toileting. Urinary tract infections should always be ruled out when there is a sudden onset of confusion [2].

Immunizations
Immunizations constitute one of the best methods of preventing common infections. Influenza vaccines are recommended for people with dementia, especially those who may not wash their hands as often as possible or those who participate in social or daycare activities with other frail people. It is also critical to have patients and families get accustomed to washing their hands throughout the day, particularly before each meal, after using the toilet, and whenever hands are soiled to prevent spreading illnesses and infections. Pneumonia vaccines should be administered as well, particularly for those in high-risk areas, including residential settings or adult day-care programs. Herpes zoster (shingles) vaccines should

be administered to avoid painful lesions in people who are not able to report early symptoms. Tetanus vaccines are also particularly important for patients with dementia, as they tend to injure themselves due to lack of safety awareness and frequently fail to treat themselves once they are injured; injuries to bare feet are particularly common. Caregivers should receive all of these vaccinations as well, and keep them up to date.

Medication concerns

Taking medications incorrectly is the most common reason for hospitalization in older adults. Due to a decreased understanding of time and increased difficulty with complicated tasks, there is an even greater hazard for a person with dementia to improperly take their medications. Even for patients with mild congnitive impairment (MCI) or mild dementia, it is strongly suggested that clinicians and/or families oversee medication usage by doing the following:

- use a dated and timed pillbox;
- have a family member fill the box and check periodically to see if medications are being taken correctly;
- firmly but gently reinforce that medications must be taken regularly;
- be aware of the patient's dietary supplements, vitamins, and other nutraceuticals that may interact with prescribed medications and cause harm.
- For people with dementia who live alone:
  - do not assume a phone call from a family member will result in taking medications correctly;
  - consider an electronic device that can dispense several doses of medications each day. Please see page 91 for useful resources; and
  - limit medications to only those that are absolutely essential, and decrease the number of times medications must be given during the day.

Nutraceuticals

Nutraceuticals can include a range of products, from herbals, vitamins, and brain-boosting supplements to prescription-strength nutrient-based products, which may claim to help memory and thinking. However, they are not as well studied as medications for AD, may have little or no benefit, interact with other medications, and are not without potential side effects. As always, a careful discussion of the risks versus the benefits of their use is the best approach for clinicians; given the concerns for the side effects that many consumers and caregivers have, a thoughtful collaboration ensures that the patient's best interests are at heart.

When medications should be stopped

In advanced stages of AD, it may not be clear how a treatment is affecting the patient. It is often only through stopping the medication that it becomes apparent what the treatment might have been preserving. In late- to end-stage AD, medications can be tapered down, resuming the medication if any negative effect is observed. Careful assessment of the need for any medications should be reviewed, particularly when the patient may be experiencing changes in swallowing and the ability to take pills.

Identifying and managing problem behaviors

AD and other diseases that cause memory loss have a variety of symptoms that can baffle and overwhelm family members. Some of the most challenging and frightening problems rarely occur early in the disease, but may happen when it is least expected as the disease progresses. Examples of such problems include the following:

- waking up in the middle of the night to get dressed and start the day;
- not recognizing familiar settings, home, or family late in the afternoon;
- becoming irritated or belligerent late in the day;
- accusing family members of stealing items the person has hidden; blaming 'outsiders' for taking things;
- accusing a spouse of infidelity;
- threatening family members with physical violence;
- refusing to bathe
- refusing to leave the house (eg, go out socially, go to doctors visits);
- walking away from home or getting lost;
- pacing back and forth without stopping;
- telling false stories;

- believing there are people in their home when there are not; and
- seeming selfish.

These behaviors may be interpreted as mean-spirited and purposeful; however, these are symptoms of the person's developing dementia and illness. These behaviors are common and expected in mid-disease and generally represent the person's fears and/or unmet needs. It is important that the person is given time and attention to express their concerns to clinicians or caregivers.

Special considerations for caregivers

The following section deals with approaches to problems that are commonly encountered when caring for people with moderate dementia. While there are no definitive answers to these problems, the approaches suggested may help.

Bathing

Caregivers often encounter issues when attempting to bathe patients with AD. Many patients with AD may refuse to bathe or say they have already bathed when they have not, which can be frustrating, especially if the person develops body odor. Many older adults are also modest about disrobing, become afraid of bath water or the shower, or are overwhelmed by the complexity of the task [10].

To help allay these fears and discomforts, caregivers should let the patient choose the time of day to bathe and how they prefer to bathe. Also, making the bathing experience more comfortable for the patient can help the process (eg, using safely secured floor mats, checking water temperature, keeping warm towels and robes readily available, using a hand-held shower head to stop the water from hitting the patient's head, allowing the person to bathe with underwear on). Bathing at the sink can also be sufficient for those resisting a shower or tub bath; using no-rinse soap in the water can simplify the process even further. Caregivers can also mentally coach the patient for bathing; for instance, complimenting the person after the bath, using distractions (eg, singing, playing music, providing snacks), associating bathing with pleasurable activities (eg, singing, snacking), and giving incentives to bathe (eg, patient must be clean before doing certain activities). If these techniques fail, caregivers can be encouraged to hire a health care assistant who the person may be more likely to accept.

Wearing the same clothing day after day

When a person wears the same clothes day after day, it is usually an indication that they cannot handle change, which is normal for many patients with moderate dementia. For patients with this issue, caregivers may consider purchasing several identical outfits when shopping; when the person takes off one set of soiled clothes, caregivers replace them with an identical set of clean clothing.

Hiding things

One of the most frustrating aspects of caring for people with moderate dementia is when they hide or lose things. Generally, a person may hide objects because they are concerned with theft. A caregiver must accept that things will be lost when caring for a person with dementia; however, it is important to minimize the loss of money and valuables.

Caregivers should remove valuables for safe keeping from the house whenever possible, and place jewelry and other valuables not used daily in a safe place (eg, safety deposit box). If valuable jewelry is worn daily, jewelers may be able to remove and replace valuable contents (eg, gems, stones) with less valuable items, to be returned to the patient. Valuables should not be given to a patient if they are staying at a nursing home or assisted living facility, since it is likely that the items will be lost. Caregivers may find it helpful to put alerting devices on keys so they may easily be found; making a routine of putting items, such as keys, in a designated location may help to keep the patient from hiding the objects. Making duplicates of important items is also helpful. Lastly, it is beneficial for caregivers to learn common places where patients may hide their objects (eg, under mattresses, within book pages, curtain hems, behind pictures, and mirrors).

Fear of abandonment/refusing help

As discussed previously, fear of abandonment is common in people with dementia when they are aware of their disease, and worry that they will lose the people who they depend on and love. When familiarity slips away, change becomes increasingly frightening, resulting in the person refusing help. Many persons refuse to go to adult day-care programs or allow in-home respite services; persons often become dependent on their caregivers to remember things when they lose their memory, and become nervous and upset when their caregiver is not around. This can become so severe that the caregiver is unable to have a moment alone; therefore, the caregiver should be encouraged to try adult day care. People who attend adult day programming adapt better to new circumstances, especially placement. In order to ensure a smooth transition, caregivers may want to stay with the person in day care for 1–2 days until the person becomes accustomed to the staff and routines.

It is additionally beneficial to have extra help in the home as early as possible to get the patient comfortable with having other people around (eg, a cleaning person). Caregivers should encourage other family members to participate in care on a regular basis and take the patient out whenever possible. While a person may become enraged when a service provider or family member is used for respite, caregivers must insist that they need time and space; caregivers must learn to gently reinforce that the person cannot stay alone or accompany the caregiver. Clinicians should coach the caregiver, reiterating that it may take a while for the respite worker and person to sync.

Aphasia

Aphasia, or loss of language abilities, is a common part of dementia, often starting with loss of reading comprehension. Warning signs include the following: mail starts to pile up; the person begins to pay anything that resembles a bill; or the person either stops reading the paper or cannot tell you what they have read. When the person starts to stumble over words, it is important to understand that they also have trouble understanding what is being said. Caregivers will find that talking more slowly, using simple phrases, gestures, or pointing to the object will help. Giving the person extra time to respond will be necessary. If the person begins to use words that do not make sense, often called 'word salad,' caregivers should try to find words that relate to the person's world; the person may have good understanding of the world around them, but may simply not be able to express themselves.

In moderation, it is acceptable to explore potential meanings with the person while they are experiencing aphasia, unless frustration begins to rise. If the patient becomes frustrated, distract them to another task, and try later. Caregivers should not be surprised if frustration and decreased inhibition might result in the use of swear words; however, caregivers and family members should not draw definitive conclusions based on a patient's outbursts of profanity. A consultation with a speech pathologist may be helpful to develop communication strategies.

If the person originally spoke a different language than the language spoken by the family and caregivers, it is expected they will return to that original language. If no one in the family speaks that language, an interpreter might help, although the person is usually aphasic in their primary language as well.

Voice tone and body language should always be considered while speaking to people with dementia; patients are able to interpret tone and postures that are both harsh and negative.

If the person develops slurred speech or problems swallowing, caregivers should speak to the physician immediately. The person may run the risk of aspirating food or saliva.

Confabulation

Confabulation, or making up stories, is another challenging outcome of memory loss for caregivers. When people with moderate dementia cannot remember what has happened, their brains tend to 'fill in the blanks;' the patient, thus, may create stories they believe are true. Caregivers should recognize that confabulations are not lies, but a story the brain has made up. Trying to correct the person's stories can lead to anger and frustration for the patient and the caregiver. Clinicians should remind the caregiver that as a rule, anything the patient says is fine, as long as safety is not compromised.

Repeated questions

Patients with moderate dementia ask repeated questions for several possible reasons: they cannot remember asking the question; they lack a sense of time; or the question they are asking is not what they really want to know. When patients ask questions repeatedly, most often it has to do with when or where something will happen, and sometimes these questions can become obsessive. There are several tips for caregivers when managing a patient's repetitive questions:

- never announce anything more than 24 hours in advance because it is likely to precipitate obsessive questions;
- when a question is asked more than once or twice, ask "Why are you asking?" then address the underlying concern; and
- another strategy is to write the answer on a file card and have the person carry it with them. When the question is asked, direct the person to read the card.

Wandering, pacing, and eloping

Wandering is a common issue with AD dementias, which involves repeatedly walking, almost aimlessly, and often getting into things (such as drawers, closets, and desks) and rummaging. Wandering is generally harmless unless the person attempts to leave or falls. Caregivers should consider putting away objects that may break and remove anything that might cause the person to trip, such as throw rugs or exposed electrical cords. The best way for a caregiver to stop someone from wandering is to redirect them to activities (see page 62). Also, caregivers should note that about 61% of people with dementia living at home will wander away, compared to 23% of people living in nursing homes [11]. If a patient with dementia tends to wander, it may be critical for caregivers to consider placing patients in nursing homes or assisted living facilities for the patient's own safety [12].

Pacing is considered continuous walking back and forth, usually at a brisk pace. People with dementia who continuously pace may not stop to eat, drink, or participate in activities. Those who wander and pace are at high risk for eloping (ie, leaving a specific area without the knowledge and approval of the caregiver, posing risk to the person with dementia). Caregivers should be instructed to secure doors and windows when the person is unattended and make sure the person has an ID bracelet, or GPS locator (see page 92 for useful resources) in case they get lost. Clinicians should consider treating people who pace with regularly scheduled pain medication (eg, acetaminophen), and caregivers should be recommended to have new shoes every few months for the patient, as pain will worsen with pacing. People with dementia who pace, wander, or elope tend to do well in adult day programming, which caregivers should consider.

Waking at night

Waking at night is another vexing symptom of dementia, as it can place undue strain on both patients and caregivers. Most patients with dementia who wake at night are confused, and are at higher risk for injuries from falls. There are several reasons why people with dementia wake at night and most are easily treatable, usually without medications. If the patient is overtired, caregivers should consider increasing the rest periods during the day to 30 minutes in the morning, a 60- to 90-minute nap after lunch, and, if the person is going to bed early (eg, before 9:00 PM), provide a third nap right before dinner. Also, a caregiver may develop a bedtime ritual or introduce daytime exercise for 20 minutes.

Physicians should be aware of medical conditions that may affect the patient's sleep (eg, mild congestive heart failure causing dyspnea, chronic obstructive pulmonary disease, gastroesophageal reflux disease, undertreated chronic pain, or nighttime administration of cholinesterase inhibitors [may cause vivid dreams or nightmares]) or sleep disorders such as sleep apnea or restless leg syndrome.

Hallucinations and paranoia

Usually hallucinations that occur in late-moderate dementia are misinterpretations of visual stimuli, television, mirror images, pictures, dolls, or figurines. By removing these stimuli from the environment and increasing activity, the illusions usually disappear.

In another form of illusion, the person may report seeing dead relatives, often occurring at night. Clinicians should advise the family that this is normal, and encourage the patient to talk about it. Clinicians should only treat illusions with medication if the hallucinations are disturbing or cause the person to be unsafe (eg, leave the house).

Patients with dementia can also begin to develop paranoid ideas. Caregivers should be instructed to not argue or correct the person, as the patient's brain is 'playing tricks' and the paranoid ideas, illusions, and beliefs are very real to them. Correcting the patient may imply that the caregiver does not know or care about what the patient is experiencing. Caregivers should reassure the patient that they are safe, that the problem has been or will be corrected, and that their family and caregivers care about the problem. If the agitation escalates and does not diminish, instruct the caregiver to get to a safe place and call the emergency services, as they may be in danger. Caregivers often think aggressive outbursts are an isolated incident; however, if left untreated, serious injury can result.

Aggression and violence

Aggression can occur for many reasons, including extra demand, psychosis, a non-AD presentation, and/or premorbid personality traits. While most physical aggression is relatively rare, verbal aggression or belligerence are comparatively common [13]. People with dementia can become depressed, get frustrated, and/or may not see their deficits as clearly as their caregivers. The inability to manage emotions is lost and anger and verbal aggression results in all moderate AD patients.

If verbal aggression occurs, care should be taken to diffuse the situation before the altercation escalates to physical violence. Clinicians should advise caregivers to make peace, even if it requires caregivers to suspend their notions about what is right and wrong. Additional techniques for diffusing verbal aggression include agreeing with the patient, apologizing, feigning ignorance, avoiding reasoning with the person (which could escalate the situation), or pausing and letting the person forget.

While certainly not a typical part of dementia, violence occasionally becomes a problem. When physical aggression occurs, caregivers often report mixed feelings of disbelief, embarrassment, guilt, shame, and denial. Aggression toward anyone is a crisis that must be dealt with immediately. Table 5.2 describes how clinicians should counsel caregivers if patients with dementia become physically aggressive and violent [13]. This table can be made into a handout for clinicians to give to caregivers.

Acknowledge the violence

Suspend everything until the episode passes

Do only what absolutely **has** to be done, such as getting any mood controlling medications

When a crisis presents, the caregiver must back down and stay back until help arrives

Show caution in one's own body language

Make sure to approach from the front and avoid turning away from patient

Make sure the caregiver stays between the person and the door

Give the patient plenty of space

Turn **off** the television, radio, and stop any extraneous stimuli

Talk in a measured, low, soft voice

Make sure directions are said as simple declarative sentences (eg, "Give me the knife," "Put the knife down")

Never try to take a weapon away from the patient during a confrontation; instead, back away slowly

Caregivers must understand that even after the person calms, the situation may happen again

Never let an aggressive angry person drive a car. Call the police to stop the person if necessary

There are basic principles of violence:

- Violent episodes are time-limited because of the energy expended
- If a patient is injured, the person can regroup from their injury and if there is a trigger, the violence **will** recur
- Untreated violence goes from bad to worse. It does not get better on its own. In every episode there are warnings, but often caregivers choose not to heed them, often out of disbelief that their loved one would never do anything to hurt them
- This is not a time to let family 'vote' on a solution. Much of the serious injury happens after the family told the caregiver that "it wasn't that serious"
- Get help immediately. The caregiver must not stay alone in the house with a violent person. The patient is in a panic mode and cannot be counted on to inhibit any impulses. Call 911 and wait for paramedics. The caregiver should never get into a car with a violent person. Once at the ER, the caregiver must be instructed not to minimize the symptoms of the episode. A prescription for a mood controlling medication and a psychiatry appointment for follow-up in 1–2 days should occur before the person is discharged home
- If the person calms when emergency personnel arrives, insist on an ER visit, as there is a good chance the behavior will start again
- If the caregiver lives alone with the person, make sure the caregiver has have a panic button or a cell phone to call for help
- The caregiver must also act defensively by planning an escape route and/or being prepared to lock themselves (with panic button or cell phone) away from a violent person
- If the person has violent tendencies:
  - Episodes are more likely to start at night. The caregiver needs to move to another bedroom with a door locked at night. Getting out of a shared bedroom alone could save serious injury
  - Instruct caregivers not to open the bedroom door if the person wakes them. A patient with dementia may fear for their lives and they believe that the caregiver is part of the problem
- Anything can be used as a weapon. There should be no guns in the house, not even in a locked cabinet. Fireplace pokers and knives should be stored out of sight. The aggressive person is panicked at the point of violence and may use a book, alarm clock, letter opener, or even a small table to injure or kill. Minimize potential weapons in the house and make sure to watch for anything that might be used

Table 5.2 Recommendations for caregivers when patients become physically aggressive (continues overleaf).

Once in the physician's office or ER, caregivers must be encouraged to avoid the following:

- Do not refuse mood-controlling medication for the patient
- Do not blame themselves for the episode
- Do not avoid placing the person in psychiatric hospitalization for fear of social stigma
- Realize that if it happened once, it **will** happen again

Table 5.2 Recommendations for caregivers when patients become physically aggressive (continued). AD, Alzheimer's disease; ER, emergency room. Adapted from © University of Iowa, 2004. All rights reserved. Hall et al [14].

References

1   Kim SY, Karlawish JH, Kim HM, Wall IF, Bozoki AC, Appelbaum PS. Preservation of the capacity to appoint a proxy decision maker: implications for dementia research. *Arch Gen Psychiatry.* 2011;68:214-220.

2   Hall G, Buckwalter K. Progressively lowered stress threshold: a conceptual model for care of adults with Alzheimer's disease. *Arch Psychiatr Nurs.* 1987;1:399-406.

3   D'Arruda KA. HIPAA update: administrative simplification and national standards. *AAOHN J.* 2002;50:496-498.

4   Brown LB, Ott BR. Driving and dementia: a review of the literature. *J Geriatr Psychiatry Neurol.* 2004;17:232-240.

5   Alzheimer's Society. Living alone. http://www.exeter.anglican.org/assets/downloads/ yourministry_downloads/Dementia%20Challenge/Living%20alone%20with%20dementia%20- %20some%20guidance1.pdf. Updated February, 2011. Accessed November 20, 2014.

6   Hall G, Bossen A, Specht J. Live-Alone Assessment. Iowa City, IA: University of Iowa College of Nursing; 2004.

7   Smith JC, Nielson KA, Antuono P, et al. Semantic memory functional MRI and cognitive function after exercise intervention in mild cognitive impairment. *J Alzheimers Dis.* 2013;37:197-215.

8   Littbrand H, Stenvall M, Rosendahl E. Applicability and effects of physical exercise on physical and cognitive functions and activities of daily living among people with dementia: a systematic review. *Am J Phys Med Rehabil.* 2011;90:495-518.

9   Spalletta G, Caltagirone C, Girardi P, Gianni W, Casini AR, Palmer K. The role of persistent and incident major depression on rate of cognitive deterioration in newly diagnosed Alzheimer's disease patients. *Psychiatry Res.* 2012;198:263-268.

10 Hall G, Gallagher M, Hoffman-Snyder C. Evidence-based Practice Guideline: Bathing Persons with Dementia. Gerontological Nursing Interventions Research Center Research Translational and Dissemination Core (RTDC). Iowa City, IA: The University of Iowa College of Nursing, The John A. Hertford Foundation Center of Geriatric Nursing Excellence; 2013.

11 Alzheimer's Association. 2012 Alzheimer's Disease Facts and Figures. https://www.alz.org/ downloads/facts_figures_2012.pdf. Accessed November 20, 2014.

12 Hope T, Keene J, McShane RH, Fairburn CG, Gedling K, Jacoby R. Wandering in dementia: a longitudinal study. *Int Psychogeriatr.* 2001;13:137-147.

13 Cohen-Mansfield J. Agitated behavior in persons with dementia: the relationship between type of behavior, its frequency, and its disruptiveness. *J Psychiatr Res.* 2008;43:64-69.

14 Hall GR. *As Memory Fades…*. Iowa City, IA: University of Iowa; 2008.

Chapter 6

Care issues in advanced dementia

The following behaviors are commonly reported by caregivers when describing persons with advanced dementia:

- may not recognize family members and/or usual caregivers;
- misinterprets caregiver's assistance with personal care, which results in yelling or striking out;
- appears withdrawn;
- falls down when attempting to get up without help;
- may only accept soft and sweet foods;
- loses weight if not offered food/fluids frequently;
- chokes more easily;
- sleeps in spurts throughout the day/night;
- fatigues easily resulting in agitation; and
- unmet needs may result in yelling or calling out.

Medical management in advanced dementia

Advanced Alzheimer's disease (AD) dementia is a terminal diagnosis and requires medical management for approximately 1–3 years of life or until death. Advanced AD dementia eventually causes the person to become incontinent, lose the ability to move independently, become unable to communicate in a meaningful way, and need total assistance for every aspect of daily living. Families often face inevitable placement of the patient into a care facility, as the 24/7 care demands exceed their best efforts. If health care decisions have been unaddressed, new issues will emerge that will necessitate discussion and decisions. Comfort becomes

© Springer Healthcare 2015
A. Burke et al., *Pocket Reference to Alzheimer's Disease Management*,
DOI 10.1007/978-1-910315-22-4_6

the overarching goal in caring for patients with advanced AD dementia and should guide all medical management decisions.

Common issues include pain management, weight loss, falls, and infections. By using the patient's behaviors as a guide to comfort, caregivers are often able to identify discomfort and report issues. In addition to excess disability (as described earlier), most behaviors experienced in advanced dementia stem from unmet physiologic or basic needs. The need-driven, dementia-compromised behavior model provides a structure and process for behaviors to be seen as a symptom or form of communication of a person's unmet need versus the problem.

Pain in dementia

Physical pain is common in many older adults due to arthritis, joint and bone diseases, and lack of movement. While people with mild memory problems can provide accurate reports of pain, those with moderate-to-advanced dementia have difficulty understanding and reporting their pain [1]. Behavioral and nonverbal signs (Table 6.1) often indicate that the person with dementia is hurting or in pain. In addition, pain can result in resistance to care, agitation, restlessness, and physical aggression.

Family and routine caregivers are often the first to recognize pain. Pain is more likely to be seen while daily care is being delivered, when a person gets in or out of a chair or bed, or as the day progresses.

Physicians play a key role in helping family and caregivers manage pain to promote comfort. While medical management of pain is described earlier, these nonpharmacological approaches can enhance pain relief:

- Provide a light touch, a light hand/foot massage, or reassurance to the person, all of which can be comforting.
- Provide a warm blanket or towel over a sore joint to help with relaxation.
- Utilize favorite distracting activities, such as music, TV programs, or favorite snacks (eg, ice cream).
- Frequent changes in positioning and movement, using favorite chairs and/or pillows; enhance the person's ability to move more freely.

Weight loss

Weight loss and dehydration are common problems as a person with advanced dementia can no longer feed themselves, and will eventually forget how to chew and swallow. Dysphagia is a prominent feature and will likely result in aspiration pneumonia, which is the most common cause of death. Feeding tubes can be avoided with careful hand-feeding. Oral intake can be enhanced by caregivers by:

- ensuring that the environment is quiet, allowing the person to focus on eating;
- leaving ample time (often up to 60 minutes) for meals. Often, small, frequent feedings may be more practical and better tolerated than a larger meal;
- encouraging a liberal diet of soft and sweet foods that are easily accepted by the patient; and
- providing sips of fluid at every opportunity while the person is awake.

Physical signs
Facial grimacing
Frowning
Eyes tightly closed
Restless body movement
Tense muscles/clenched fists
Rubbing, holding, or guarding a body part
Pulling away or hitting when touched
Increased pacing
Wanting to exit the home
Verbal signs
Moaning
Grunting
Crying out
Repeating words or phrases
Behavioral signs
Crying
Acting fearful
Increased agitation
Change in normal sleep pattern
Decreased appetite

Table 6.1 Common pain behaviors. Adapted from © Elsevier, 2009. All rights reserved. Scherder et al [1].

Falls

Falls are increasingly common as the person loses strength, coordination of movement, spatial abilities, and lack of insight on the need for increased assistance. If a person with increased risk of falling is restrained, unwanted behaviors of physical and verbal agitation are likely to appear [2]. The following strategies can provide a source of comfort and may help minimize falls:

- Encourage caregivers to anticipate the person's need for movement and assist the person throughout the day in sit-to-stand motions and/or simple short steps.
- Consult a physical therapist to work with caregivers on a range of motion, movement, and positioning programs.
- Ensure that the person wears good footwear with non-skid soles.
- Remind caregivers that despite their efforts, falls will happen in advanced dementia. The focus of care is to prevent injury.

Infection

Since infection is a common cause of death in patients with dementia, families will need to decide if there is a point where treatment of an infection should cease. Managing any symptoms related to the infection and keeping the person comfortable is imperative. Dyspnea, usually due to pneumonia, is common at end of life. The use of oxygen, antipyretics, and analgesia should be consistently employed.

Palliative care approaches

Many families of persons with advanced dementia usually say that their loved one would choose comfort (ie, palliation) and would not want their life prolonged by medical interventions [3]. Therefore, they may choose to have 'do not resuscitate' (DNR) directives, refuse feeding tubes, and decline hospitalization.

When discussing advanced dementia care with families, clinicians should not say: "What do you want for your loved one?" but instead say, "What would your loved one ask for if they could tell us? Would they ask for comfort care or would they want their life prolonged by medical means?" This wording may help take the guilt and burden off the family.

If a family has chosen palliative or comfort care for the patient, then the clinician should consider:

- no antibiotics, except for comfort: discuss antibiotics with the family. Comfort indicates no intravenous or intramuscular antibiotics, as injections are uncomfortable. If the family says a person would not want to be "alive like this," then oral antibiotics should also be discontinued; there are a few exceptions that can be decided at the time, for example, orchitis and parotitis;
- no laboratory testing;
- no dietary limits;
- no blood glucose monitoring: if the family wants blood sugars followed, then keeping blood sugars less than 300 mmol/L will prevent hyperosmolar coma. Tighter controls are not needed for this population, and one blood glucose test per month is adequate;
- no appetite stimulants: there are no data indicating that any medication (including megace and cyproheptadine) is effective in stimulating appetite in persons with advanced dementia;
- no daily or weekly blood pressure checks;
- no measuring food intake and output; and
- simplify medications: no cholesterol-lowering drugs, no aspirin to prevent stroke, no tight blood pressure control, and minimized cardiac medications.

Hospice and dementia

Hospice care focuses on comfort and support rather than cure or rehabilitation. Families are generally in favor of comfort care in dementia; however, they usually do not know who to ask or know when it is time to receive hospice care.

The following guidelines are helpful to know when the person may be eligible for hospice care [4]:

- Life expectancy is 6 months or less if the disease runs its natural course.
- The person is unable to walk, bathe, and dress independently.
- There is no meaningful verbal communication.
- One or more of the following has occurred in the past year:

- aspiration pneumonia;
- kidney/urinary tract infection;
- recurring fever after antibiotics;
- pressure ulcers (bed sores); and/or
- weight loss.

Because the course of dementia is so unpredictable, a person in the earlier stages of dementia (ie, still walking and speaking well) who continues to lose weight despite caregiver efforts may qualify for hospice. There may be other life-limiting conditions, such as heart or lung disease, diabetes, stroke, and cancer that may also allow them to qualify.

Hospice services comprise all products and services related to the terminal diagnosis of dementia, including:

- the hospice team: physician, nurse, social worker, nursing assistant, chaplain, dietician, pharmacist, and volunteers;
- care that is individualized and based on the person and family's goals of comfort and quality of life;
- supplies, equipment, and medications;
- service where the person resides;
- in-home services for management of any acute symptoms;
- respite care for caregiver relief, which provides temporary admission for up to 5 days in a skilled nursing facility or hospice palliative care unit; and
- bereavement counseling following death.

Possibility of abuse

As judgment and reasoning become more impaired, the person with AD becomes a vulnerable adult who is more at risk for exploitation. The person's increasing repetitiveness, dependence, and possible behavioral changes can be very hard for even the best of caregivers, and the risk is real for unwitting or serious abuse. In addition, as the patient's ability to manage household-related functions, such as cooking, paying bills, taking medications, and driving becomes more impaired, their safety and overall welfare can become significantly affected if appropriate levels of support have not been put in place. Families may inadvertently endanger their loved ones by not increasing supervision and assistance because they may minimize the person's symptoms or remain long-distance care providers. The person may become at risk of financial scams and/or detrimental family or outside influences. Physically, the patient may present with labile blood pressure, significant weight loss, depression, and even frank signs and symptoms of abuse. With increasing need for assistance with activities of daily living, the patient may appear more disheveled, wear inappropriate or stained clothing, and have body odors. As neurologic impairment becomes more pronounced they are more at risk for falls, incontinence, and recurrent infections, all of which can have secondary signs of neglect or abuse from care providers.

References

1   Scherder E, Herr K, Pickering G, Gibson S, Benedetti F, Lautenbacher S. Pain in dementia. *Pain*. 2009;145:276-278.

2   Evans LK, Cotter VT. Avoiding restraints in patients with dementia: understanding, prevention, and management are the keys. *Am J Nurs*. 2008;108:40-49.

3   Mitchell SL, Teno JM, Kiely DK, et al. The clinical course of advanced dementia. *N Engl J Med*. 2009;361:1529-1538.

4   Ross JS, Sanchez-Reilly S. Hospice eligibility card. http://geriatrics.uthscsa.edu/tools/Hospice_elegibility_card__Ross_and_Sanchez_Reilly_2008.pdf. Accessed November 20, 2014.

Chapter 7

Supporting caregiver health

Caregiver responsibilities and needs

Having discussed the role of the caregiver in supporting a patient with Alzheimer's disease (AD) dementia in previous chapters, it is important to explain briefly what we understand by the term 'caregiver.' No definitive explanation of the term exists; however it is generally accepted that a caregiver provides extraordinary help and care, exceeding the bounds of what is normative or usual, sometimes at a great cost to themselves. There is also a general consensus that this support is provided predominantly by a partner or family member [1].

For a caregiver, it may seem that the whole focus is on the care, supervision, and quality-of-life maintenance for the person with dementia. Caregivers may feel like life has become so focused on meeting the needs of the person with AD dementia that their needs become the lowest priority, which can be difficult to experience. Yet, the support provided by the caregiver is invaluable for the person with AD dementia.

Research shows that caregiving is a very complex activity; it is draining yet rewarding; demanding yet fulfilling; boring yet challenging; and intimate yet lonely. Caregivers often suffer from guilt, fatigue, and depression, as their responsibilities may span over decades. The effects of caregiving can also manifest as physical health issues as caregivers may neglect their own health and suffer from stress-related illnesses. This increased risk of physical and mental health problems, can in turn result in an increased risk of death for the caregiver. Thus, being a caregiver who is experiencing mental or emotional strain is a risk factor for

© Springer Healthcare 2015
A. Burke et al., *Pocket Reference to Alzheimer's Disease Management*,
DOI 10.1007/978-1-910315-22-4_7

mortality [1,2]. Stresses may be compounded when caregivers continue other responsibilities as well, such as raising children or grandchildren, or being employed. Moreover, families can often be conflicted about how caregiving should be accomplished or how much participation should be expected from each member.

Caregiving responsibilities do not end when the person with dementia is admitted to a care setting. Families and caregivers must monitor care continuously, balancing expectations with the realities of long-term care environments, providing certain aspects of care, and advocating for the patient. The clinician should consider the needs of both patient and caregiver, suggesting rest and services to maintain the caregiver's endurance.

Taking care

Caregivers often experience a difficult time planning for their own needs. As the symptoms of their loved one's illness become more pronounced and more demanding on their time, caregivers often neglect social relationships and their physical and emotional health. Moreover, friends and family members may expect this degree of self-sacrifice as a part of marriage vows or their perception of familial obligations, putting extra pressure on the caregiver.

For adult children to support caregivers or become caregivers themselves, they may need help fully understanding dementia and what they can do to help. All caregivers, whether that is a partner or family relation, will need coaching to voice their needs and to generate a written list of care needs, which can be utilized at a family meeting to open negotiations (Table 7.1).

It is critical for a caregiver to remember that often they may need to ask for help or support themselves. Caregivers must be their own vocal advocate in meeting personal needs; they should know and understand that this is not selfish, it is survival. Please see page 95 in the Appendix for the 'Taking care of yourself' handout, which may be copied and given to caregivers to support their health and well-being.

From a distance	Living nearby
Send funny greeting cards or flowers regularly	Pick up dry cleaning
Review sources of information on the internet and forward it to them	Take dinner to them once a week
Call at least twice a month and listen – even if it is repetitive or sad; tell the caregiver and person with AD dementia how much you care	Call or visit at least twice a month; when you visit, just listen
	Help with cleaning and fix-ups
	Take the person with dementia out for a meal
Send news of your children and pictures	Stay with the person while the caregiver gets a haircut
Plan to visit at least 2–3 times a year and stay in a hotel	
Plan to attend diagnostic appointments	Serve as a sounding board, without feeling like you have to 'fix' problems
Remember all birthdays, anniversaries, and holidays	Take the caregiver to a support group
Consult with other family members on specific needs	Each month monitor the person with dementia's driving
Send favorite foods occasionally	Help to find local resources, including legal resources
Send theater tickets or perhaps a subscription to a film rental service	Help with legal and financial issues as appropriate
Keep up to date on loved one's dementia	Take the caregiver and person with dementia to religious services
Compliment the caregiver on a job well done; compliment nearby relatives on the help they provide	Stay with person so caregiver can go to a favorite social group
Plan to provide actual onsite respite 1–2 times a year for 1–2 days at a time	

Table 7.1 Ideas for helping the caregiver. AD, Alzheimer's disease.

Family conflict

Caregivers should expect family conflict while caring for a person with dementia. Conflict can be caused by the stress experienced from caregiving, the burden of the responsibility of care, or the pressure of possibly having to take control of a person's assets if required. Families may argue as a way of coping with the long-term stress of caring for a loved one with dementia (written communication, W Carron, August 1998). Each member will go through stages of grief at their own pace. Family members should understand that fighting needs to be fair and a family therapist should be consulted if needed.

Support groups

Clinicians should refer caregivers to local support groups (eg, Alzheimer's Association) or provide them with take-home educational material. Many people find support groups very helpful, while others may feel embarrassed by them. Remind the caregiver that it may take a few tries before they find a group they like, but it is usually worth the effort.

Online support

For those who have access to a computer, there are numerous informational websites for neurological diseases, AD, and caregiver support groups. Please see page 91 for useful resources on online support groups. Additionally, suggest that caregivers search online to find appropriate support groups and information for them and their situation.

References

1 Schulz R. Martire LM. Family caregiving of persons with dementia: prevalence, health effects, and support strategies. *Am J Geriatr Psychiatry*. 2004;12:240-249.
2 Schulz R, Beach SR. Caregiving as a risk factor for mortality: the Caregiver Health Effects Study. *JAMA*. 1999;282:2215-2219.

Chapter 8

Appendix

Useful resources

The following recommended resources contain further information about
Alzheimer's disease (AD) and related topics.

Websites

There are numerous sites for information on AD and related disorders;
some are listed below. However, families and caregivers should be warned
of websites with misinformation and/or websites that offer products and
advertised treatments or even cures for AD, especially when the product
claims to be a dietary supplement, an ancient treatment, or the producer
implies that there is a medical conspiracy to hide information. When an
official advance is made in treating AD, clinicians can access that information on the following sites; these sites have additional information on
cutting-edge research data and supportive information for caregivers and
family members:

The Alzheimer's Association

www.alz.org

A comprehensive site developed to allow families and professionals to
locate nearby Alzheimer's Association chapters and resources. The site
also provides up-to-date research, library of reference materials, and
caregiver information.

© Springer Healthcare 2015
A. Burke et al., *Pocket Reference to Alzheimer's Disease Management*,
DOI 10.1007/978-1-910315-22-4_8

The Alzheimer's Association: Safety center

www.alz.org/care/dementia-medic-alert-safe-return.asp

This page within the Alzheimer's Association website provides family members and caregivers access to helpful programs and information for managing a patient with AD. This includes information on medical identification bracelets and a 24-hour nationwide emergency response service for individuals with AD or a related dementia who wander, become lost, or have a medical emergency. Through this program, the Alzheimer's Association will provide 24-hour, nationwide assistance, no matter when or where the person is reported missing. Identification jewelry associated with this program is available through this webpage.

The Alzheimer's Association Location-Based Mapping Service

www.alz.org/comfortzone/

The Alzheimer's Association can provide devices that give location updates of a person with AD, so family members can monitor that person's location, while the individual with AD can maintain their independence and enjoy the emotional security of familiar routines and surroundings.

Virtual Library at the Alzheimer's Association Green-Field Library

www.alz.org/library/index.asp

The Green-Field Library provides a comprehensive list of books relating to AD and related topics, including educational and personal stories, caregiving issues, professional instruction, and much more.

Alzheimer's Store

www.alzstore.com

An online resource providing access to the following websites for caregivers and professionals, managing a patient with AD and/or related dementias.

US National Institutes of Health, Clinical Trials Information

www.clinicaltrials.gov

Clinicaltrials.gov provides information to consumers and clinicians about the medications currently being studied for AD, the location of study sites, type of studies, the status of the study, and whether or not the sites are recruiting subjects. To use the site, simply type 'Alzheimer's disease' in the search box.

National Institute on Neurological Disorders and Stroke (NINDS) Alzheimer's Disease Information Page

www.ninds.nih.gov/disorders/alzheimersdisease
Published by the US Department of Health and Human Services and the National Institute on Neurological Disorders and Stroke, this site provides up-to-date information on dementia, its related illnesses, and invaluable links to reputable organizations and information.

The National Institute on Aging (NIA) Alzheimer's Disease Education and Referral (ADEAR) Center

www.nia.nih.gov/Alzheimers/
The Alzheimer's Disease Education and Referral (ADEAR) Center website helps users find current, comprehensive AD information, and resources from the National Institute on Aging (NIA).

The Alzheimer's Disease Fact Sheet

www.nia.nih.gov/Alzheimers/publication/alzheimers-disease-fact-sheet
One service offered by ADEAR is the Alzheimer's Disease Fact Sheet. This publication can be copied and distributed to friends and family who do not understand the basics of dementia.

Books

There are many books available to families coping with AD dementia. Some popular guides for families are available at most bookstores or online book sellers including the following:

- Powell L, Courtice K. *Alzheimer's Disease: A Guide for Families*. 3rd edition. Cambridge, MA: Perseus Publishing; 2002.
- Bell V, Troxell D. *A Dignified Life, Revised and Expanded: The Best Friends Approach to Alzheimer's Care: A Guide for Care Partners*. Deerfield Beach, FL: HCI Books; 2012.

- Callone P, Kudlacek C, Vasiloff BC, Manternach J, Brumback RA. *A Caregiver's Guide to Alzheimer's Disease: 300 Tips for Making Life Easier.* New York, NY: Demos Medical Publishing, LLC; 2006.
- Warner M. *In Search of the Alzheimer Wanderer: A Workbook to Protect Your Loved One.* West Lafayette, IN: Purdue University Press; 2005.

Many of the caregiver/family guides were written in the early 2000s; however, their advice is still germane today.

Reporting abuse

The toll-free number for the National Center on Elder Abuse (Eldercare), which has information on locating state agencies where you can report physical, emotional, and sexual abuse including financial exploitation and neglect, is 1-800-677-1116. A state-by-state resource directory is available online at www.ncea.aoa.gov/NCEAroot/Main_Site/Find_Help/State_Resources.aspx. Other international resources include the International Network for the Prevention of Elder Abuse: http://www.inpea.net/ and the dementia information pages on the website of AgeUK: http://www.ageuk.org.uk/health-wellbeing/conditions-illnesses/dementia/.

Taking care of yourself: a handout for caregivers

Taking care of yourself is the most important thing you can do to take care of your loved one. Some suggestions include:

1. Eat right: make sure you get a balanced diet
2. Get adequate rest
3. Drink plenty of fluids: drinking 1.5–2 quarts per day of any fluids (not strictly water) is suggested for good health. (Many people with AD will drink sweetened drinks, milk, tea, or coffee but refuse water)
4. Exercise at least 3 times a week (5 times a week is preferable) for a minimum of 20 minutes. Some exercise ideas include:
    *   Walking, especially with a good friend; mall-walking; shopping
    *   Dancing with your loved one
    *   Gardening
    *   Bowling
    *   Swimming
    *   Exercising with a pet
5. If you find that you are having the following symptoms for 2 weeks or more, call your doctor and ask about treatment for depression. There is NO SHAME in being depressed; depression is a common problem for caregivers. Check the following symptoms if you are feeling them, and tell your doctor:
    ☐ Low mood, continuing sadness, crying
    ☐ Feelings of hopelessness or despair
    ☐ Feelings of guilt
    ☐ Changes in appetite and/or weight loss
    ☐ Changes in sleep patterns
    ☐ Inertia, inability to get things started or completed
    ☐ Nervousness or anxiety
    ☐ Irritability
    ☐ Memory loss or confusion
    ☐ Unexplained aches and pains anywhere
6. Make sure you get annual health screenings
7. Get flu and pneumonia vaccinations
8. Make sure your tetanus immunization is current
9. Get out with friends and by yourself regularly
10. Make sure you have some time alone each day
11. Pursue a hobby, especially with family or friends
12. Find something that makes you laugh and do it regularly
13. Rent movies you enjoy
14. Attend your spiritual or religious center regularly
15. Use respite services, especially adult day programming as often as possible; 3 days per week is recommended
16. Attend support groups and keep in touch with professionals
17. Do something mentally challenging (eg, crosswords or puzzles)
18. Splurge on yourself (eg, shopping)
19. Always remember that you are doing something special for someone you care deeply about; that makes you very special
20. Talk to friends, family members or other people that you trust, making sure to ask for help if or when it is needed.